SpringerBriefs in Energy

SpringerBriefs in Energy presents concise summaries of cutting-edge research and practical applications in all aspects of Energy. Featuring compact volumes of 50 to 125 pages, the series covers a range of content from professional to academic. Typical topics might include:

- A snapshot of a hot or emerging topic
- A contextual literature review
- A timely report of state-of-the art analytical techniques
- An in-depth case study
- A presentation of core concepts that students must understand in order to make independent contributions.

Briefs allow authors to present their ideas and readers to absorb them with minimal time investment.

Briefs will be published as part of Springer's eBook collection, with millions of users worldwide. In addition, Briefs will be available for individual print and electronic purchase. Briefs are characterized by fast, global electronic dissemination, standard publishing contracts, easy-to-use manuscript preparation and formatting guidelines, and expedited production schedules. We aim for publication 8–12 weeks after acceptance.

Both solicited and unsolicited manuscripts are considered for publication in this series. Briefs can also arise from the scale up of a planned chapter. Instead of simply contributing to an edited volume, the author gets an authored book with the space necessary to provide more data, fundamentals and background on the subject, methodology, future outlook, etc.

SpringerBriefs in Energy contains a distinct subseries focusing on Energy Analysis and edited by Charles Hall, State University of New York. Books for this subseries will emphasize quantitative accounting of energy use and availability, including the potential and limitations of new technologies in terms of energy returned on energy invested. The second distinct subseries connected to Springer-Briefs in Energy, entitled Computational Modeling of Energy Systems, is edited by Thomas Nagel, and Haibing Shao, Helmholtz Centre for Environmental Research - UFZ, Leipzig, Germany. This sub-series publishes titles focusing on the role that computer-aided engineering (CAE) plays in advancing various engineering sectors, particularly in the context of transforming energy systems towards renewable sources, decentralized landscapes, and smart grids.

All Springer brief titles should undergo standard single-blind peer-review to ensure high scientific quality by at least two experts in the field.

Amrutha Mary Varkey · Eby Johny

Energy Economics and Policy

Navigating the Global Energy Market

 Springer

Amrutha Mary Varkey
School of The Environment
Coventry University
Coventry, UK

Eby Johny
Faculty of Political Science
and International Studies
University of Warsaw
Warsaw, Poland

ISSN 2191-5520   ISSN 2191-5539 (electronic)
SpringerBriefs in Energy
ISBN 978-3-032-05481-4   ISBN 978-3-032-05482-1 (eBook)
https://doi.org/10.1007/978-3-032-05482-1

© The Author(s), under exclusive license to Springer Nature Switzerland AG 2025

This work is subject to copyright. All rights are solely and exclusively licensed by the Publisher, whether the whole or part of the material is concerned, specifically the rights of translation, reprinting, reuse of illustrations, recitation, broadcasting, reproduction on microfilms or in any other physical way, and transmission or information storage and retrieval, electronic adaptation, computer software, or by similar or dissimilar methodology now known or hereafter developed.
The use of general descriptive names, registered names, trademarks, service marks, etc. in this publication does not imply, even in the absence of a specific statement, that such names are exempt from the relevant protective laws and regulations and therefore free for general use.
The publisher, the authors and the editors are safe to assume that the advice and information in this book are believed to be true and accurate at the date of publication. Neither the publisher nor the authors or the editors give a warranty, expressed or implied, with respect to the material contained herein or for any errors or omissions that may have been made. The publisher remains neutral with regard to jurisdictional claims in published maps and institutional affiliations.

This Springer imprint is published by the registered company Springer Nature Switzerland AG
The registered company address is: Gewerbestrasse 11, 6330 Cham, Switzerland

If disposing of this product, please recycle the paper.

*To our fathers, **Johny** and **Varkey**, and our mothers, **Meritta** and **Alice**, in gratitude for their enduring love.*

Preface

Climate change is the effect of the greatest market failure and a negative global production externality. While classical economics emphasizes free-market mechanisms and capitalist incentives in resource allocation, neoclassical economics has not fully succeeded in providing solutions to modern environmental challenges such as climate change and greenhouse gas emissions. The energy sector contributes to global emissions to a great extent. The energy sector is the most direct source of carbon emissions. Fossil fuels dominate the energy demand globally; owing to their limited reserves available and ability to emit harmful gases, they still dominate the consumption mix for a long period of time. This poses a major challenge in addressing climate change. This book highlights modern-day solutions to environmental problems in the context of energy transition and discusses the market mechanisms adopted since the Kyoto Protocol, which played a crucial role in mitigating global environmental challenges, particularly through the growth of the international carbon market since the 1990s.

This book is intended for undergraduates, postgraduates, scholars, professionals, and environmental practitioners. It introduces climate change as a global negative production externality, war and carbon emissions, the social cost of carbon emissions, carbon pricing, the carbon tax and ETS, the geopolitics of oil, tariffs, and trade wars with microeconomic foundations. Those with limited backgrounds in economics can benefit from the simple illustration of the economics of climate change, fossil fuels, and tariffs. This book addresses the geopolitical aspects of energy markets and the volatility of energy prices. We recommend this book as a quick guide to gain insights into energy economics, climate change, energy pricing, the geopolitics of energy, organizational carbon accounting, carbon footprint, and energy policy.

This book covers not only the aspects of climate change, negative externalities, carbon pricing, and the energy crisis but also highlights carbon accounting, carbon footprint calculations and the need for renewable energy transitions. Community renewable energy projects play a crucial role in the transitioning energy scenario. Sustainable livelihoods are the main concern for stakeholders, governments, and the public. This book highlights the Hockerton Housing Project, an example of a sustainable community and livelihood, and the Hornsea offshore wind farm, which is the largest wind farm in the world. Sustainable livelihoods and practices for renewable energy adoption increase energy efficiency, reduce energy costs, and reduce carbon emissions. This book illustrates the recent trends in energy production and the changing energy mix. It is suitable for students of Energy and Environmental Economics, Climate Economics, Oil and Gas Management, and Renewable Energy Management.

Energy production and consumption reveal excessive market failures and externalities that the free market fails to address. This book takes a critical stance toward free-market environmentalism and the necessity of policy interventions. Traditional approaches to energy economics should be recontextualized beyond free-market philosophies and environmental neocolonialism. This book draws insights from my lectures on Energy Economics and Climate Change at Coventry University, UK. It has been further enriched through collaboration with Eby Johny, an expert in economics and geopolitics, who teaches Geopolitics of Aid at the University of Warsaw, Poland, whose perspectives on global production, externalities, and geopolitics of fossil fuels have added depth to the analysis.

The first four chapters of this book discuss climate change as a negative global production externality, the economics of fossil fuels, trends in energy production and a changing energy mix, and the geopolitics of oil, energy, and trade wars. Chapter 5 focuses on carbon accounting and footprint calculations, and Chap. 6 focus on renewable energy transition, and energy policies.

Coventry, UK   Amrutha Mary Varkey
                                                                                       Eby Johny

Competing Interests The authors have no competing interests to declare that are relevant to the content of this manuscript.

Contents

1 **Climate Change as a Negative Global Production Externality** 1
   1.1   Introduction 1
   1.2   Social Cost of War Emissions 3
   1.3   Global Production and CO_2 Emissions from Fossil Fuels 5
   1.4   Monetary Price on Carbon Emissions 5
   References 8

2 **Economics of Fossil Fuels** 11
   2.1   Global Energy Demand and Supply 11
   2.2   Energy Crisis in Europe 13
   2.3   Crude Oil Export Trends 15
   References 18

3 **Geopolitics of Oil and Trade War** 21
   3.1   Geopolitical Interest in Oil 21
       3.1.1   Venezuela 22
       3.1.2   Iraq and the Gulf War 22
       3.1.3   Libya 24
       3.1.4   Yemen Conflicts 24
       3.1.5   Sudan and South Sudan 25
       3.1.6   Nigeria 25
       3.1.7   Democratic Republic of the Congo and Republic
             of the Congo 26
       3.1.8   Angola 27
       3.1.9   Gabon, Equatorial Guinea, and Cameroon 27
       3.1.10 The Central African CFA Franc Currency and the Oil
             Market 27
   3.2   Oil- and Mineral-Based Neoliberal Institutions for Central
       African Countries. 28
       3.2.1   Petrodollars and GCC Countries 28
   3.3   Oil as a Geopolitical Weapon 29
       3.3.1   Soviet Union Versus Afghanistan 29

		3.3.2	Ukraine and the Russian War	30
		3.3.3	America and Great Britain	30
		3.3.4	NoN-OPEC	31
		3.3.5	OPEC Behavior and Strategy	31
		3.3.6	OPEC +	32
	3.4	Energy and Trade War		32
		3.4.1	Trade Tactics After the Cold War	33
		3.4.2	Effect of the U.S. Tariff on China	33
	References			36

4 Recent Trends in Energy Production and the Changing Energy Mix ... 39
- 4.1 UK's Trends in Energy Production from 2000 to 2022 ... 40
- 4.2 Poland's Trends in Energy Production from 2000 to 2022 ... 44
- 4.3 India's Trend in Energy Production from 2000 to 2022 ... 46
- References ... 50

5 Carbon Accounting and Footprint Calculation ... 51
- 5.1 Introduction ... 51
- 5.2 Why Carbon Accounting ... 52
- 5.3 SSE Energy Solutions Carbon Footprint ... 54
- 5.4 Apple's Strategies for Carbon Footprint Reduction ... 56
- References ... 58

6 Renewable Energy Transition and Energy Policies ... 61
- 6.1 Hockerton Housing Project ... 62
- 6.2 Hornsea Offshore Wind Farm Projects (Hornsea 1 and 2) ... 63
- 6.3 UK's Carbon Footprint ... 64
- 6.4 Energy Policies and Recent Developments ... 66
- 6.5 India's Energy Policy ... 67
- 6.6 Politics of Renewable Energy ... 69
- References ... 70

Index ... 73

About the Authors

Amrutha Mary Varkey is a lecturer, Economics in Coventry University, UK. Her research interests are Energy Economics, Climate Economics, Carbon Accounting, Sustainable Innovation, and Urban Development. She has been a researcher at the University of Warsaw since 2021 and taught Development Economics at the Faculty of Economic Sciences, UW. She engaged in the FIRMINREG—Gospostrateg VI research project in the Faculty of Economic Sciences and the Ministry of Finance, Poland (grant from the National Centre for Research and Development), 2021–2022. She was an affiliate at Adam Smith Business School, University of Glasgow (2022–2023). She was a Research Fellow at the Institute for Global Food Security, Queen's University Belfast, and worked on NERC-funded project on Biodiversity and Green Finance (2023–2024). She is the author of book *The Rural to Urban Transition in Developing Countries: Urbanisation and Peri-Urban Land Markets* (Routledge, 2023).

Eby Johny is a researcher and teacher of Geopolitics of Aid in University of Warsaw, Poland. His research interests are Foreign Aid Diplomacy, Geopolitics of Energy, Trade War, South Asian Relations, Secular Democracy in India, and Federalism. He has a profound interest in international affairs, Foreign Aid Diplomacy, the Global South, Trade, and the Environment. Eby Johny has been a researcher at the faculty of Political Science and International Relations, University of Warsaw, since 2021.

List of Figures

Fig. 1.1	Annual CO_2 emissions. *Source* Author's illustration via OWID	3
Fig. 1.2	The quantity of production of war equipment, illustration by the author	4
Fig. 1.3	Per capita CO_2 emissions by fuel type, 2022. *Data Source* Per capita CO_2 emissions by fuel type, 2022 based on Global Carbon Budget, 2023, OWID, illustration by the author	6
Fig. 1.4	Price trend for the EU ETS. This graph shows the price trend of the EU ETS. *Source* Illustration by the author using World Bank data	7
Fig. 2.1	Total energy supply. *Source* Total energy supply with IEA, 2021 data, illustration by the author	12
Fig. 2.2	The world's total energy supply. *Source* Compiled by the author via the IEA, 2021	12
Fig. 2.3	Brent Crude Oil prices during 2014–2024. *Source* Author's compiled using a 10-year daily chart with macrotrend data	14
Fig. 2.4	Crude oil export trend from 2000 to 2022. [11–15] *Source* Compiled by the author	16
Fig. 2.5	Crude oil export trend from 2000 to 2022. [11–15] *Source* Compiled by the author using	16
Fig. 3.1	Strategic locations. *Source* Google	23
Fig. 3.2	Distribution of crude oil reserves in Central Asia as of 2021. *Source* Author's illustration using African Energy Commission, Statista 2024	26
Fig. 3.3	Effects of tariffs on Chinese production Quantity of Chinese goods exported to the U.S. (x-axis) and price (y-axis)	34
Fig. 4.1	Coal, peat, and oil shale in the UK during 2000–2022	41
Fig. 4.2	Crude, NGL, and feedstocks in the UK during 2000–2022	41
Fig. 4.3	Oil, UK	42
Fig. 4.4	Natural gas, UK	43
Fig. 4.5	Nuclear energy in the UK	43

Fig. 4.6	UK's trends in renewables and waste	44
Fig. 4.7	Coal, peat, and oil shale in Poland	44
Fig. 4.8	Oil products in Poland	45
Fig. 4.9	Natural gas in Poland	45
Fig. 4.10	Renewables in Poland	46
Fig. 4.11	Coal, peat, and oil shale in India	47
Fig. 4.12	Crude, NGL, and feedstocks in India	47
Fig. 4.13	Oil products in India	48
Fig. 4.14	Natural gas	48
Fig. 5.1	Word cloud of key terms associated with carbon accounting	53
Fig. 6.1	Field visit, Hockerton housing project, Nottinghamshire	63
Fig. 6.2	Major emitting industries on a residence basis: UK, 1990 to 2023. Greenhouse gas emissions, UK: provisional estimates from the Office for National Statistics (ONS) (in million tonnes of CO2e)	65

Chapter 1
Climate Change as a Negative Global Production Externality

Abstract Climate change and global warming are the most significant negative externalities affecting global production. Atmospheric concentrations of carbon dioxide and other greenhouse gas emissions have increased remarkably since preindustrial levels. The global increase in carbon dioxide is due to increasing fossil fuel use and land-use changes. This chapter discusses climate change as a negative externality of global production, the social cost of war emissions, and the monetary price of carbon emissions. The increase in global temperature and carbon emissions is attributed to the high demand for energy for residential, industrial, and commercial purposes. An increase in global energy demand drives global warming and results in changes in climate over time. Since the Industrial Revolution, human activities have had atmospheric concentrations of greenhouse gases on the Earth's surface. The climate has experienced long-term changes in temperature, weather, sea-level rise, and precipitation. This introductory chapter addresses public goods, market failure, externalities of war, and carbon emissions, and how imposing a price on carbon can reduce global production externalities. Human activities had already warmed the planet by about 1°C above pre-industrial levels, and achieving the Paris Agreement goal of limiting global warming to 1.5°C is now increasingly challenging. Excessive carbon emissions need to be tracked and controlled to avoid great risks to humanity and the economy. In this context, this chapter highlights carbon pricing as an instrument to control carbon emissions.

Keywords Climate change · Global negative production externalities · War · Emission · Social cost of carbon · Public goods

1.1 Introduction

Climate change is a negative externality associated with widespread damage to global public goods. The reasons why humans fail to address global production externalities and mitigate market failures are the quest for profit, limitless growth, and less capital to invest in clean energy. CO_2 emission is the major contributor to global warming.

Households, businesses, and nations contribute to the emissions and spillovers of the externalities of economic growth. Increasing the social cost of GHG emissions makes traditional bargaining solutions infeasible [23]. The theory of public good by Paul Samuelson is the cornerstone of environmental economics. Before the 1960s, A. C. Pigou suggested a method of limiting pollution by taxing polluters for each unit of emissions. In the 1960s, Ronald Coase proposed a new form of intervention through bargaining solutions [2]. However, the ground-breaking solutions by Coase, Crocker, and Dales failed to set limits on the environment [3, 5, 6], neoclassical economics did not address the modern challenges humanity faced by setting limits to environmental problems [10]. Undoubtedly, the Kyoto Protocol had a limited capacity to slow CO_2 emissions. The burning of fossil fuels results in high CO_2 emissions, leading to significant negative externalities. Global production leading to high demand for energy drives negative global production externalities.

Climate change is the effect of the greatest market failure [1, 22], it is a negative global production externality. A varying climate is an economic and political dilemma resulting from the inability to avoid a freeriding environment [8, 28, 29]. The environment is a global public good. These types of global environmental damage range from groundwater depletion by multinational companies, emissions from war, emissions from power generation, innovation externalities, and several channels of global production. In today's context, the old challenges have disappeared, and the negative production externalities are more global than local and geopolitical, as the challenge is quite different from that of previous centuries, as, today, businesses and geopolitical problems aggravate public bads (negative externalities). For example, an excessive quantity of emissions is from the Russia–Ukraine war and the Palestine War. War contributes to carbon emissions; the number of global public bads are increasing. War equipment uses more fossil fuel, and war leaves behind a greater carbon footprint. Humans leave behind more carbon footprints, and organizations do the same. We live in an illusion of a free market economy, where the market resolves its own problems, but today's problem is not just market failure but also government failure.

An increase in demand leads to an increase in global production, which manifests in greater amounts of public bads and externalities. Global production decreased during the COVID-19 pandemic, followed by demand, consumption, global production externalities, and emissions from factories and production. However, the extent of global production externalities has been on the rise since the war began in late 2021, as Russia–Ukraine tensions started, with increased amounts of negative production externalities accruing from emissions from war, defense, and military expenses amounting to a massive carbon war footprint caused by falling bombs, resulting in smoke and fire damaging infrastructures. Restructuring "Gaza" requires a greater amount of energy and, therefore, great emissions and externalities. During times of conflict or war, the demand for arms increases due to the anarchy structure of global politics. It contributes to the wealth of massive arms-producing countries or private entities that produce arms. Public bads are on the rise with the increase in global emissions of CO_2.

1.2 Social Cost of War Emissions

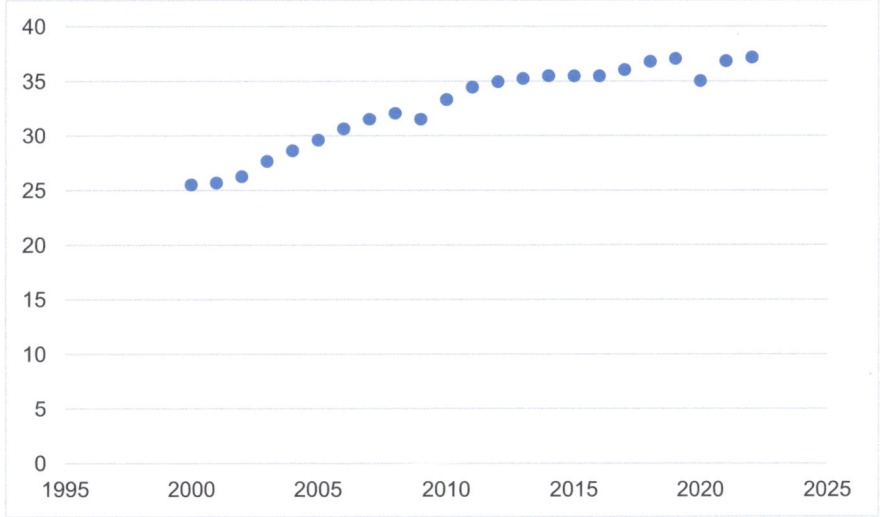

Fig. 1.1 Annual CO_2 emissions. *Source* Author's illustration via OWID

The historic increase in CO_2 emissions is shown in Fig. 1.1 as the demand for global energy increases. In 2022, emissions accounted for 37.15 billion metric tons of CO_2. There was a dip in CO_2 emissions during COVID-19, 2019–2020, due to the decrease in global production levels and industrial activities, and less demand in sectors such as aviation, and a further increase in emissions during 2022 between Russia and Ukraine aggravated the global energy demand and resulted in the energy crisis. Fossil fuel emissions are the major driver of CO_2 emissions. Rising CO_2 emissions have clear environmental consequences, and excessive emissions are associated with the negative externalities associated with production.

1.2 Social Cost of War Emissions

Significant carbon emissions results from conflicts in "Ukraine" and "Gaza." War is a negative externality of production through excess defense production. The externality associated with war leads to the loss of biodiversity; reconstruction efforts result in a greater number of negative externalities. War carbon emissions are enormous but remain unaccounted for. Social cost is associated with the damage resulting from war emissions. During times of war, there is an increase in the production of war equipment, resulting in goods with negative externalities. Demand for arms increases with the anticipation of a conflict or war. The social cost of carbon is greater if the emissions are greater. The social cost of carbon can be avoided by reducing carbon dioxide emissions [24, 25].

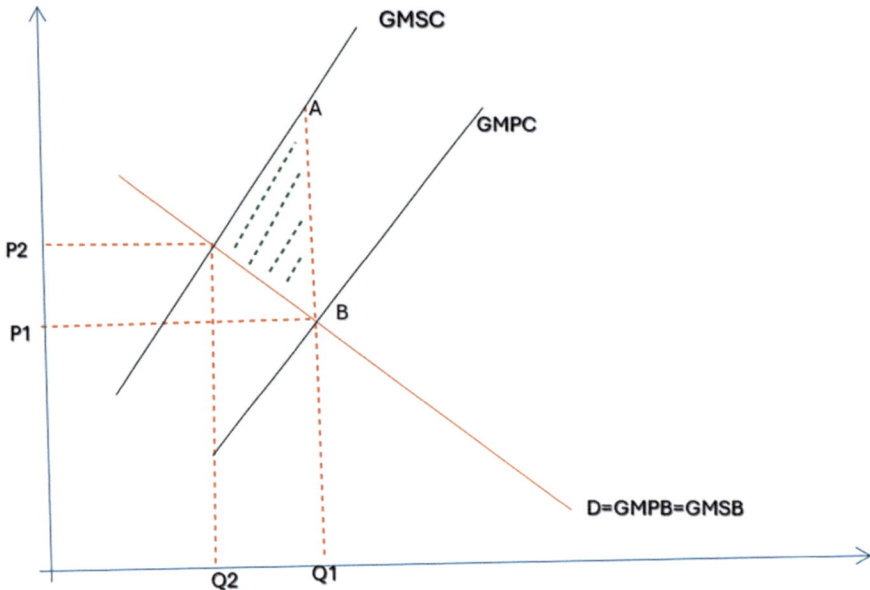

Fig. 1.2 The quantity of production of war equipment, illustration by the author

Figure 1.2 social cost of the global negative production externality of war by the author; GMSC refers to the global marginal social cost, GMPC is the global marginal private cost, and GMPB is the global marginal private benefit. The quantity of production of war equipment, illustration by the author.

The social cost of war refers to adverse effects; it includes direct and indirect socioeconomic and environmental costs to society. These costs are largely borne by society. During the war, the global marginal social cost increased; it added to social costs from additional war-related activities and the global production of arms. The social cost of war is the summation of the individual marginal cost of the production of war equipment associated with the war. The private marginal cost refers to the direct cost of bombing, including fuel and personnel. The destruction of homes, civilian death, environmental damage, terror, and fear are the external marginal costs of the war. GMPC, the global marginal private cost, captures the externalities or spillover effects of the global aggregate production of war equipment [GMPC = $MPC_1 + MPC_2 + MPC_3 + \ldots + MPC_n$]; see Fig. 1.2.

$$GMSC = GMPC + \text{External Costs}$$

The demand for defense goods increases during war or in anticipation of war. Arms dealers foresee the global scenario of conflicts and tensions while enhancing the production of such equipment. A global levy on war production equipment has many benefits, as it reduces production and controls the supply of arms. Negotiations and reducing the war production of equipment reduce social costs and emissions.

The social cost of the global negative production externality of war creates costs for society and the environment. War requires more fossil fuel, thereby resulting in negative externalities. The global production of arms and industrial production results in increased emissions via the burning of fossil fuels.

1.3 Global Production and CO_2 Emissions from Fossil Fuels

Global energy demand and industrial production generate high CO_2 emissions through the burning of fossil fuels such as coal, oil, and natural gas. Globally, industrialization has led to the use of fossil fuels. Fossil fuels constitute the greatest source of electricity production worldwide. Energy accounts for two-thirds of global emissions [21]. Emissions are low in Latin America and the Caribbean. Africa has considerable emissions from cement and flaring. Emissions from oil are high in South America and North America.

The highest amount of CO_2 emissions is from coal [11], the second highest is from oil, the third highest is from natural gas, and the next highest is from cement and flaring (see Fig. 1.3). In the early centuries, coal production was dominant in association with industrialization in countries such as the UK and the USA [4, 7]. In recent decades, Asian countries such as India and China have had high production and consumption of coal, and Africa also has the highest industrial production. In India, there are 1.854258 billion metric tons of annual emissions from coal, and in China, there are 8.250736 billion metric tons. Emissions from oil and natural gas are highest in the United States, with 2.249519 billion metric tons. Although developing countries such as India are blamed for high emissions, their per capita emissions are lower than those of developed countries.

1.4 Monetary Price on Carbon Emissions

Global industrial production leads to additional social costs of carbon. The social cost of carbon (SCC) refers to the measurement of economic damage due to the release of one ton of carbon dioxide (CO_2), or it is the monetary value of damage by an extra ton of CO_2. The SCC is the key tool used in climate change policy [12, 13, 14, 16, 17, 26, 27]. The increasing emissions from industrial production increase the social cost of carbon. Governments and companies use the social cost of carbon to set a price or carbon tax to reduce emissions.

The social cost of carbon (SCC) represents the monetary cost associated with climate damage that results from the emission of an additional tonne of carbon dioxide (tCO_2). The global SCC (GSCC) accounts for the externality of CO_2 release and is thus the right value to use from a global welfare perspective [18]. Climate

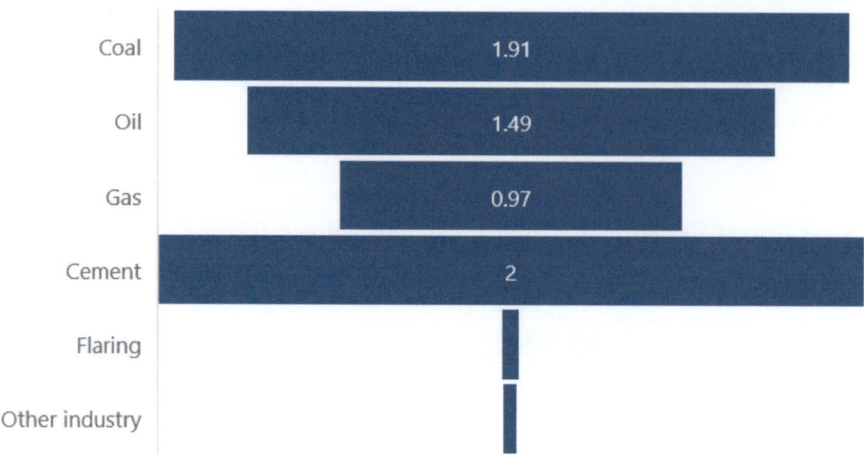

Fig. 1.3 Per capita CO_2 emissions by fuel type, 2022. *Data Source* Per capita CO_2 emissions by fuel type, 2022 based on Global Carbon Budget, 2023, OWID, illustration by the author

change is not the only negative externality of the burning of fossil fuels; it poses a great threat to people's lives. Putting a price on carbon emissions via a carbon tax may have a positive effect on emission reduction, as it shifts production and consumption from carbon-intensive goods to low-carbon sources.

Global emissions rise with income and industrial development. Carbon pricing is a policy targeted at addressing consumption choices. The two categories of carbon pricing include emissions trading systems (ETSs) and carbon taxes [9]. The ETS is a cap-and-trade scheme that sets a cap on total greenhouse gas emissions, allowing businesses with high emissions to buy allowances (for example, the EU ETS). The ETS sets a market price for greenhouse gas emissions; see (Fig. 1.4).

The carbon tax sets a monetary value on carbon emissions or the carbon content of fossil fuels [30]. Figure 1.4 shows that in 2021, approximately 6% of emissions were in countries with a carbon tax, and 20% were covered by a trading system. In total, a carbon price was imposed on 26% of global emissions. As per the agreed-upon emission targets of the Kyoto Protocol, most countries are unable to meet their emission targets. The EU ETS is a known example of a cap-and-trade mechanism; however, as with emissions targets, the outcome is not encouraging. Both the carbon tax and ETS are similar in economic terms; they both increase the price of carbon.

Climate change is largely driven by individual behavior, industrial emissions, and high energy consumption. Industrial emissions and fossil fuel exploration affect the global climate, leading to extreme weather impacts. Often, trading market mechanisms are not successful in curbing the external costs of environmental damage and industry-led global production, demanding government intervention via a carbon tax and cap-and-trade mechanisms [15]. Market-based instruments urge polluters to pay to damage the environment by reducing greenhouse gas emissions [21].

1.4 Monetary Price on Carbon Emissions

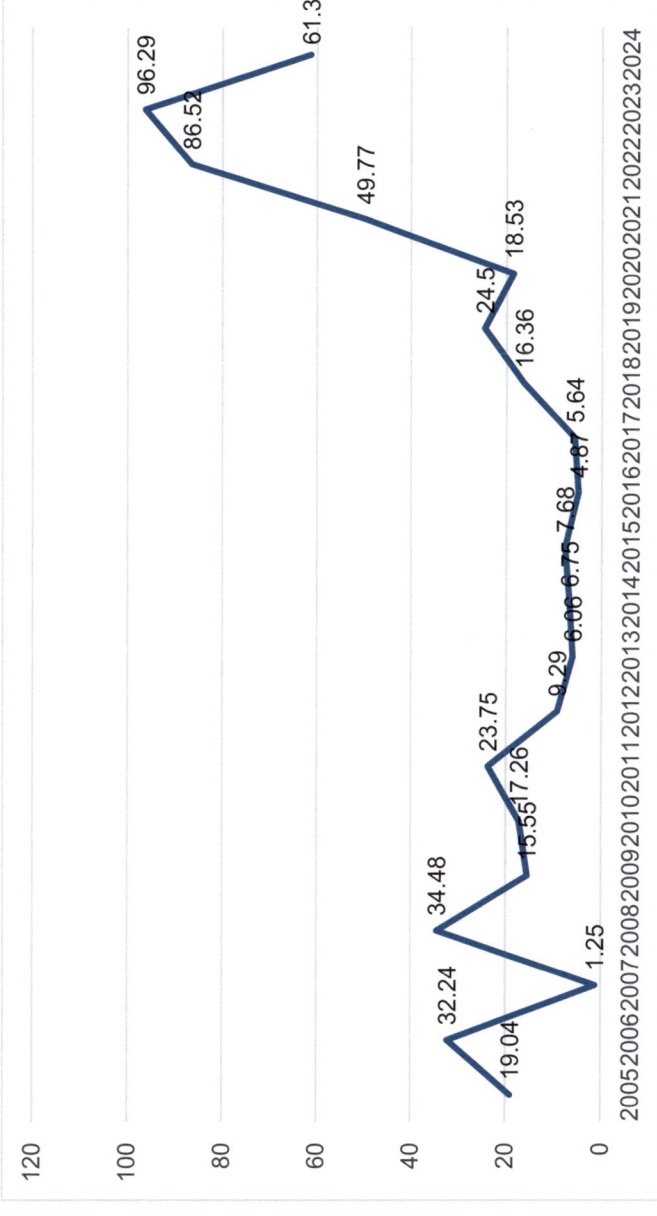

Fig. 1.4 Price trend for the EU ETS. This graph shows the price trend of the EU ETS. *Source* Illustration by the author using World Bank data

Global emissions have a devastating impact on people's lives. They are not just localized externalities but also cause ultimate challenges for humanity. For example, atomic energy is a less expensive source of energy, but the externalities resulting from accidents involving nuclear reactors and power plants will take generations to overcome the negative externalities on the environment and the lives of people. The Fukushima (2011) and Chornobyl (1986) nuclear disasters led to externalities with long-term impacts on health and the environment. The radioactive leakage into the Pacific Ocean of Fukushima increased social costs, where the private benefit of nuclear energy production outweighs the benefits to society [19].

Summary

Climate change is a major global production externality that affects people and societies, and human-induced pollution and greenhouse gas emissions aggravate the climate crisis. The long-term macroeconomic policy framework of every country should focus on resolving the potentially harmful effects of climate change and production externalities. The energy sector contributes to global emissions and production externalities to a great extent. Climate policy interventions are needed to address the problem of externalities of production. Market mechanisms and low-carbon energy technologies are long-term options for reducing greenhouse gas emissions. When a developed country emits tons of CO_2 per year, it does not internalize the negative externalities of production. Developed nations, often presuppose that developing nations will embrace green transitions quickly, overlooking their basic challenges. For a developing nation, the transition to green energy incurs high costs and takes time due to the lack of innovations and sufficient capital. Blaming the developing world in the name of environmentalism is a new form of environmental imperialism. In a free market economy, there is a massive number of competitors for the same product; the consumer has a wide variety of choices for the same product. As the market offers extensive choices, it exploits resources to increase the supply of final goods, further increasing emissions. Thus, global production augments the damages to the global public good, i.e., the environment. Further discussions on the economic analysis of climate change and the impact of the energy market highlight the negative externalities of fossil fuel production.

References

1. Andrew B (2008) Market failure, government failure and externalities in climate change mitigation: the case for a carbon tax. Public Admin Dev Int J Manag Res Pract 28(5):393–401
2. Coase RH (1959) The federal communications commission. J Law Econ 2:1–40
3. Coase RH (2013) The problem of social cost. J Law Econ 56(4):837–877
4. Committee on Climate Change (2020) Reducing UK emissions: 2020 progress report to parliament. theccc.org.uk
5. Crocker TD (1966) The structuring of atmospheric pollution control systems. Econ Air Pollut 61:81–84
6. Dales JH (1968) Land, water, and ownership. Can J Econ/Rev Can Econ 1(4):791–804
7. Department for Business, Energy & Industrial Strategy (2020) UK greenhouse gas emissions, 1990 to 2019. gov.uk

References

8. Dietz T, Shwom RL, Whitley CT (2020) Climate change and society. Ann Rev Sociol 46(1):135–158
9. Gerlagh R, Kverndokk S, Rosendahl KE (2009) Optimal timing of climate change policy: interaction between carbon taxes and innovation externalities. Environ Resource Econ 43:369–390
10. Hills J, Welford R (2005) Coca-Cola and water in India. Corpor Soc Responsibil Environ Manag 12(3)
11. IEA (2023) Coal 2023. IEA, Paris. https://www.iea.org/reports/coal-2023. License: CC BY 4.0
12. Nordhaus WD (2017) Revisiting the social cost of carbon. Proc Natl Acad Sci 114(7):1518–1523
13. Pearce D (2003) The social cost of carbon and its policy implications. Oxf Rev Econ Policy 19(3):362–384
14. Pezzey JC (2019) Why the social cost of carbon will always be disputed. Wiley Interdiscipl Rev Clim Change 10(1):e558
15. Pigou AC (1962) The economics of welfare, 4th edn. MacMillan & Co., London
16. Pindyck RS (2019) The social cost of carbon revisited. J Environ Econ Manag 94:140–160
17. Pizer W, Adler M, Aldy J, Anthoff D, Cropper M, Gillingham K, Wiener J (2014) Using and improving the social cost of carbon. Science 346(6214):1189–1190
18. Ricke K, Drouet L, Caldeira K, Tavoni M (2018) Country-level social cost of carbon. Nat Clim Chang 8(10):895–900
19. Steinhauser G, Brandl A, Johnson TE (2014) Comparison of the Chernobyl and Fukushima nuclear accidents: a review of the environmental impacts. Sci Total Environ 470:800–817
20. Stern N (2006) The Stern review on the economics of climate change. HM Treasury
21. Stern N (2007) The economics of climate change: the Stern review. Cambridge University Press
22. Stern N (2008) The economics of climate change. Am Econ Rev 98(2):1–37
23. Tirole J (2008) Some economics of global warming. Riv Polit Econ 98(6):9–42
24. Tol RS (2011) The social cost of carbon. Annu Rev Resour Econ 3(1):419–443
25. Tol RS (2023) Social cost of carbon estimates have increased over time. Nat Clim Chang 13(6):532–536
26. UK Parliament (2008) Climate change act 2008. legislation.gov.uk
27. UNFCCC (2015) The Paris agreement. unfccc.int
28. UNFCCC (2011) COP26 reaches consensus on key actions to address climate change. UN Climate Press RELEASE, 2021. https://unfccc.int/news/cop26-reaches-consensus-on-key-actions-to-address-climate-change. Accessed on 13 June 2022
29. Weitzman ML (2016) How a minimum carbon price commitment might help to internalize the global warming externality (No. w22197). National Bureau of Economic Research
30. Wolozin H (1966) Economics of air pollution

Chapter 2
Economics of Fossil Fuels

Abstract The price of essential commodities is subject to demand–supply dynamics and market mechanisms. The reasons for the increasing price of energy commodities often cannot be explained by demand–supply forces but rather by geopolitical factors. The increasing population drove the global energy demand, industrialization of developing economies, and increased the need for transportation and industrial production worldwide. This chapter discusses the global energy demand and supply and addresses the historical scenario of the oil crisis and the twin energy crisis from 2021 to 2022. The energy demand fell sharply as transport, business, and economies slowed, and Europe relied upon Russia for natural gas. The Russian–Ukrainian conflict impacted the energy distribution networks in Europe. The energy crisis during 2021–2022 disrupted global supply chains and led to supply shocks and spiraling prices. Fossil fuel production causes more than 90% of global externalities worldwide. In the fossil fuel category, oil is in the highest demand. Excessive reliance on fossil fuels increases CO_2 emissions, affects the global climate, and has many negative production externalities.

Keywords Demand–supply dynamics · Energy markets · Oil price · Russia–Ukraine war · Crude oil export · Energy crisis

2.1 Global Energy Demand and Supply

Energy is an essential commodity and fundamental to all human activities. The worldwide demand for energy has increased the demand for fossil fuels. Global energy use has increased from 8588.9 (Mtoe) in 1995 to 13,147.3 Mtoe in 2015 [10]. The total primary energy consumption of fossil fuels has recently reached nearly 80% [2, 4, 16, 23]. Asia is estimated to significantly shift energy use in non-OECD countries. The pressures from the twin global energy crisis have decreased, but global energy markets are volatile and unsettled, and fuel prices are down from their peak levels in 2022. Globally, the energy sector, particularly fossil fuel production, causes more than 90% of air pollution [11]. Energy commodities such as gasoline, diesel, natural gas, crude oil, coal, and electricity are used for various residential requirements, such as cooking, heating, and industrial and transportation [21, 22]. The world's total supply of fossil fuels has been increasing (see Fig. 2.2) in the last two decades. Among the fossil fuel category, the highest demand is for oil, coal, and natural gas. Oil demand has increased by 18 mb/d over the past two decades. Oil constituted 29% of the total supply worldwide; coal constituted 27%, whereas natural gas constituted 24% in 2021 (see Fig. 2.1).

High oil demand arises with the increasing demand for road transport. Road transport accounts for 45% of the global oil demand and is the second highest in the petrochemical sector at 15%. Gasoline demand peaked in 2017, 2018, and

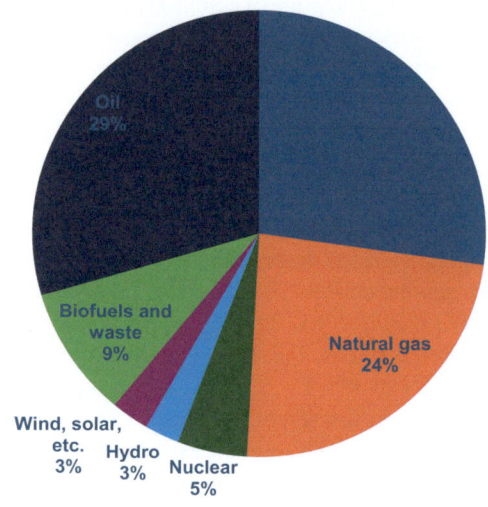

Fig. 2.1 Total energy supply. *Source* Total energy supply with IEA, 2021 data, illustration by the author

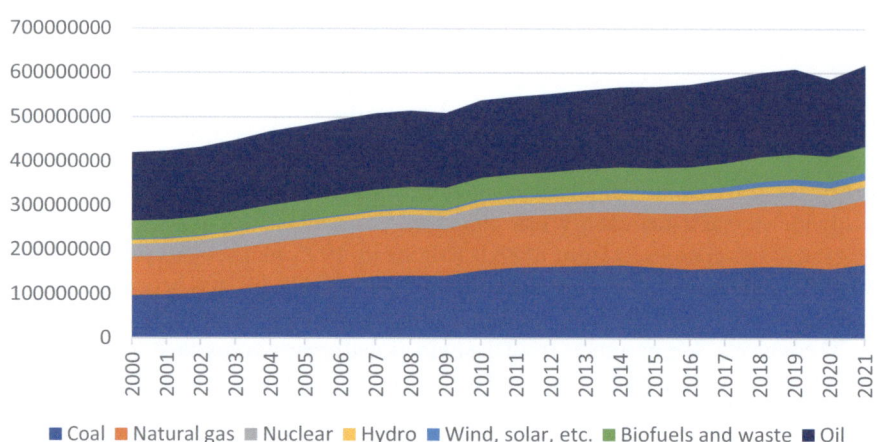

Fig. 2.2 The world's total energy supply. *Source* Compiled by the author via the IEA, 2021

2019; further changes in gasoline demand slowed during the pandemic. However, oil demand for the petrochemical and transportation sectors, such as the aviation and shipping sectors, will continue to increase by 2050. Coal is the second-largest energy source and the world's largest electricity generation source [11–15]. Coal is used for the production of steel, cement, and power. The initial demand for coal in the 2000s was driven to a large extent by China's rapid industrialization. The proportion of coal-fired power in worldwide capacity additions was nearly 45% in 2006; in developed countries, coal demand increased in 2007 and was reduced to 11% in 2022. China is the world's largest producer and consumer of coal. India is the second-highest producer, and the U.S. is the third-highest producer. Changes in iron and steel production lead to a decrease in coal demand. There has been a shift toward renewable energy resources and a decline in coal usage since early 2020 [11–15].

The developing world has a massive demand for coal production and consumption. Coal is the second-largest form of energy. The global demand for coal fell during the pandemic. However, the demand for coal remained high in 2022 despite the energy crisis. The global demand for coal is high in Asian countries such as India and China. China is the largest coal producer and is often the most readily available energy resource. The coal demand fell considerably in all advanced countries in Europe and the UK by 2023. The IEA report reveals that coal demand increased in China in 2023 (by 220 Mt, or 4.9%), followed by India (by 98 Mt, or 8%), and Indonesia (up 23 Mt, or 11%). China accounts for 54% of global coal consumption [11–15]. Developed countries are committed to the renewable energy transition by phasing out coal. The global demand for natural gas increased by an average of 2% in 2011, but research has shown a decrease in the use of 0.4% from recent years to 2030. Natural gas is commonly used in the power and building sectors. The demand for natural gas has peaked because of heating, residential, and industrial demand. The building sector is a primary energy-consuming sector that is estimated to account for more than a third of the world's energy consumption [18].

2.2 Energy Crisis in Europe

The twin crisis resulted in supply shocks and an energy crisis. The shortage of fossil fuel reserves caused the 2021–2022 crisis because of the halt in oil production during the COVID-19 pandemic, and the Ukraine and Russia conflict [17, 19, 20]. The COVID-19 pandemic significantly impacted global energy demand, as countries locked down people and businesses restricted their activities to contain the spread of the virus. In March 2020, global commercial flights were reduced by 80%, and road traffic levels were reduced by more than 50% in many countries compared with 12 months previously [7]. The causes of the oil energy crisis are attributed to disruptions in the supply chain, shortages of fossil fuel, and demand–supply mismatches. Since late 2019 (at the beginning of the pandemic), there has been a considerable decrease in energy consumption and, hence, in energy prices across the globe.

In early 2022, there was a rising demand for energy commodities, as Europe faced a shortage of natural gas, due to Russia's reduced supply to Europe. Prices increased because of the increasing military activity of Russia near the border of Ukraine, as Russia was using most of its resources for war-related purposes. The political conditions and Ukraine's effort to be a part of the EU and NATO aggravated the Russian Invasion of Ukraine [5, 6]. The continuous war and terror disrupted the energy supply in Europe and affected the global energy supply chain [19]. Europe depended on Russia for natural gas, and Europe was a significant source of Russia's revenue. The Russian–Ukraine conflict added more problems to the energy crisis. The Russian invasion of Ukraine is mainly political in nature and affects the energy supply, leading to rising prices. To distribute to European energy distribution networks, energy companies cannot source energy from Russia's natural gas supply. European energy companies relied on Russia for energy imports. European regions were heavily dependent on Russia for importing natural gas.

For the first time globally, oil prices increased in the 1970s. An oil embargo sets restrictions on the supply of oil. Historically, 1973–1974 and 1978–1979 were known for their supply shocks and rise in oil prices. Historically, an increase in oil prices was responsible for recessions, inflation, and low levels of economic activity [3]. The Arab Oil Embargo in 1973 set restrictions on the export of oil to the U.S. The oil price hikes in the 1970s marked a crisis in political history. Ten years ago, the Brent crude price was $67.31 (2014 average). According to the latest trend in November 2024, the crude oil price is $74.89. The price peak was 122.29 in March 2022 and escalated during the Russia–Ukraine war (see Fig. 2.3).

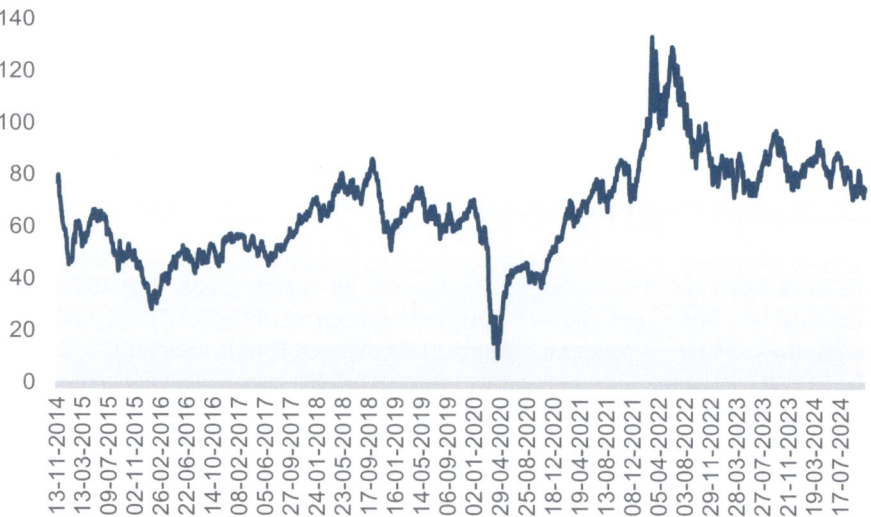

Fig. 2.3 Brent Crude Oil prices during 2014–2024. *Source* Author's compiled using a 10-year daily chart with macrotrend data

Global production showed a different trend after COVID-19. To a great extent, it is influenced by the regulatory role that OPEC plays; non-OPEC production accounts for 60% of the global production. The OPEC price in 2023 was 82.95 $ per barrel. OPEC comprises 11 countries: Algeria, Angola, Congo, Equatorial Guinea, Gabon, Iraq, Iran, Kuwait, Libya, Nigeria, Saudi Arabia, Venezuela, and the United Arab Emirates. Saudi Arabia has a major stake in oil reserves. The significant changes in global production during the pandemic affected prices. The decrease in prices from 2020 to 2021 is due mainly to the decrease in transportation demand. OPEC is critical for stabilizing volatile market conditions, such as the pandemic or the COVID-19 crisis. Member countries are the monopolies of oil production. Saudi Arabia holds a unique position in determining oil prices. During the time of the Ukraine war, the crude oil price spiked to 110 $ per barrel. OPEC provides flexibility to oil prices during wartime [1].

The price of gasoline rose in most European countries from October 2021 until March 2022. This was due to the rising demand for gasoline when fossil fuel energy reserves in Europe were significantly depleted and during the conflict between Ukraine and Russia from November 2021 to March 2022, when Russia invaded Ukraine [20]. The war restricted the production and supply of energy products. However, the recovery after the 2021–2022 crisis was quicker than after the 2007–2008 financial crisis. Many lessons were learned from this crisis, which resulted from historical and recent energy crises. Countries focused on self-sufficiency and domestic production of oil rather than export dependency. For the first time, the risk of depending on fossil fuels was identified in the context of geopolitical vulnerabilities.

2.3 Crude Oil Export Trends

Figure 2.4 shows the crude oil export trends of the regions from 2000 to 2022. The five leading exporters are the Middle East, OECD Americas, OECD Europe, Russia, and Africa. The Middle East has a significant share of exports with 22,590.63 (b/d), the second highest in OECD Americas with 14,575.88, the third major exporter is OECD Europe with 9047.54, and Russia and Africa are fourth and fifth with 70,757.21 and 5662.84 (b/d), respectively.

Global population growth and the industrialization of developing economies, particularly in China, India, and Africa, drove a steady increase in global energy demand during 1994–2019. The world's crude oil and petroleum product exports vary across the years. Notably, these countries include the USA, Saudi Arabia, Russia, and Iraq. The world's largest crude oil exporter is the United States, and Saudi Arabia is the second-largest exporter. Saudi Arabia's crude oil exports were 8,338.4 b/d in 2019, which decreased to 7,675.3 b/d in 2020 and 7,571.1 b/d in 2021. Russia's crude oil exports decreased from 7034.27 (b/d) to 6875.64 (b/d) (see Fig. 2.5 and Table 2.1).

The energy crisis of 2021–2022 affected crude oil production because of the two crises of the COVID-19 pandemic and the Russia–Ukraine war. Notable exporters

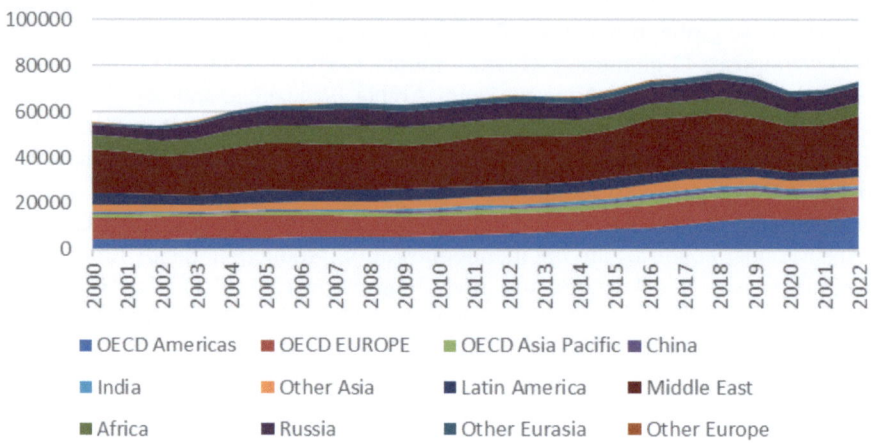

Fig. 2.4 Crude oil export trend from 2000 to 2022. [11–15] *Source* Compiled by the author

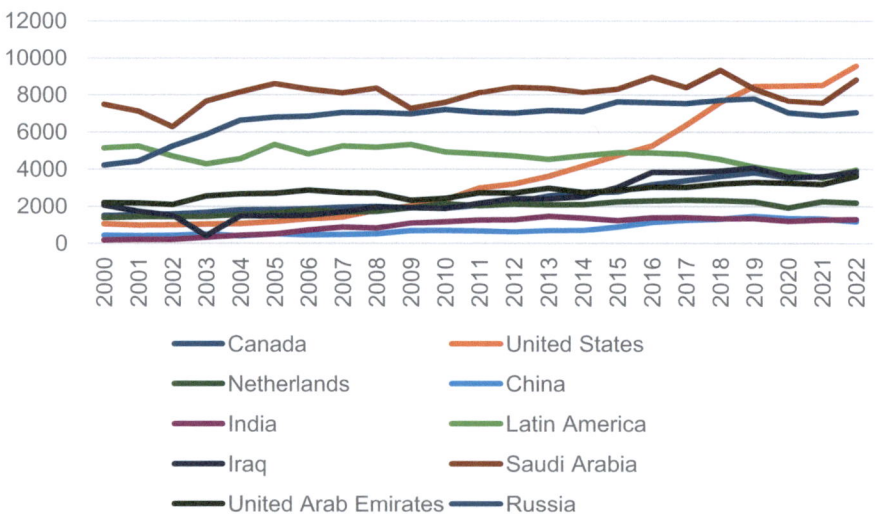

Fig. 2.5 Crude oil export trend from 2000 to 2022. [11–15] *Source* Compiled by the author using

are Venezuela, Kazakhstan, and Iran. The export levels over the last ten years have varied due to geopolitical and economic factors. The United States, the Netherlands, and India play crucial roles in the export of petroleum products. The U.S. increased exports of 9577 million in 2022 from 3205 million in 2012, showing its rising stake in the global energy market. Canada showed an increasing trend of increasing exports, from 2225 million in 2012 to 3764 million in 2022, with close ties with the U.S.

2.3 Crude Oil Export Trends

Table 2.1 Crude oil export trend during 2012–2022

	2012	2013	2014	2015	2016	2017	2018	2019	2020	2021	2022
Canada	2225.129	2563.831	2740.132	2785.154	3177.439	3384.054	3621.742	3814.494	3548.004	3618.295	3763.983
United States	3205	3621	4176	4738.667	5261	6376	7601	8471	8498	8536	9577
Netherlands	2123.746	2090.278	2093.551	2251.214	2310.172	2340.938	2314.553	2255.185	1925.328	2263.9	2193.218
China	618.51	697.8429	704.8862	893.6686	1147.054	1265.786	1328.465	1476.094	1356.697	1341.63	1189.039
India	1288.708	1475.076	1374.083	1245.4	1401.7	1411.351	1344.288	1359.78	1216.294	1283.582	1312.107
Latin America	4718.164	4535.051	4723.339	4887.001	4877.997	4803.323	4534.004	4129.058	3855.538	3537.688	3972.547
Iraq	2425.529	2401.88	2524.835	3018.584	3840.301	3822.111	3907.67	4074.855	3579.419	3597.68	3884.49
Saudi Arabia	8418.85	8364.69	8141.294	8318.057	8965.946	8406.476	9342.743	8338.381	7675.257	7571.112	8832.02
United Arab Emirates	2721.014	2981.637	2752.588	2856.246	3037.848	3032.399	3212.51	3295.988	3257.938	3195.809	3620.022
Russia	7016.30	7157.62	7099.14	7625.07	7581	7536.36	7717.05	7795.22	7034.27	6875.64	7057.21

Source Compiled by the author [11–15]

The Netherlands experienced fluctuations in its oil trade; exports were 2251 million in 2015 and declined to 2193 million in 2022. China's exports showed rapid growth as the manufacturing potential increased, and exports varied from 618.51 million in 2012 to 1189 million in 2022. This growth has decelerated in recent years due to trade tensions in the global economy. India's exports showed moderate variations and reached 1312 million in 2022, although it lags behind other economies.

Iraq faced significant fluctuations, and exports increased to 3884.49 million in 2022. Saudi Arabia is the lead player, with 8000 million exports sustained by the petroleum industry. The export trend suggests a complex landscape of global trade. The long-term results prove that Saudi Arabia tries to satisfy the world's crude oil demand by increasing production. When the world's crude oil demand increases, Saudi Arabia exploits higher prices with larger volumes, leaving almost half of the increased demand for the remaining producers (production sharing strategy). Finally, crude oil prices are more sensitive to only OECD member production in the long run than OPEC production, whose member is Saudi Arabia. This makes Saudi Arabia pursue more multilateral decisions, as a different approach would decrease its production share in a low-price environment [8, 9].

Summary

This chapter examines the scenario of global energy demand and supply, indicating how geopolitical factors influence demand–supply dynamics, and explores rising energy demand with the increase in the worldwide population and industrialization. This chapter analyzes the historical energy crisis and the twin energy crisis of 2021–2022. The energy crisis of 2021–2022 affected crude oil production because of the two crises of the COVID-19 pandemic and the Russia–Ukraine war. This chapter discusses the crude oil price trends and crude oil export trends of lead players such as Saudi Arabia and other OPEC members. Crude oil exports are influenced by supply-side geopolitics and excessive energy demand. Further discussion of the book addresses the geopolitics of oil, energy, and trade wars.

References

1. Ali H, Shadrach B, Jaimu NJ (2022) Role of OPEC+ on oil prices in the Russia-Ukraine invasion. Int J Account Res 7(2):131–137
2. Argus Media (2023) (price database). https://direct.argusmedia.com/
3. Barsky RB, Kilian L (2002) Oil and the macroeconomy since the 1970s. J Econ Perspect 18(4):115–134
4. BGR (Bundesanstalt für Geowissenschaften und Rohstoffe) (2021) (German Federal Institute for Geosciences and Natural Resources 2021). Energiestudie 2021. Daten und Entwicklungen der Deutschen und Globalen Energieversorgung. [Energy Study 2021. Data and Developments in German and Global Energy Supply]. https://www.bgr.bund.de/DE/Themen/Energie/Downloads/energiestudie_2021.pdf?__blob=publicationFile&v=4
5. Choonara J (2022) The devastation of Ukraine: NATO, Russia and imperialism. Int Soc 2(174):3–30

References

6. Cohen A (2022) Can energy exports save The Russian war machine facing Western Sanctions? Forbes. www.forbes.com/sites/arielcohen/2022/03/18/can-energy-exports-save-the-russian-war-machine-facing-western-sanctions
7. Cook M (2021) Trends in global energy supply and demand. In: Developments in petroleum science, vol 71. Elsevier, pp 15–42
8. Dagoumas A, Perifanis T, Polemis M (2017) An econometric model to assess the Saudi Arabia crude oil strategy. https://core.ac.uk/download/565339944.pdf
9. Dagoumas A, Perifanis T, Polemis M (2018) An econometric analysis of Saudi Arabia's crude oil strategy. Resour Policy 59:265–273
10. Dong K, Dong X, Jiang Q (2020) How does renewable energy consumption lower global CO_2 emissions? Evidence from countries with different income levels. World Econ 43(6):1665–1698
11. IEA (2023e), World energy outlook 2023. IEA, Paris. https://www.iea.org/reports/world-energy-outlook-2023, Licence: CC BY 4.0 (report); CC BY NC SA 4.0 (Annex A)
12. IEA (2023a) Renewable energy market update. https://www.iea.org/reports/renewable-energy-market-update-june-2023
13. IEA (2023b) SDG 7: data and projections. https://www.iea.org/reports/sdg7-data-and-projections
14. IEA (2023c) World energy balances. https://www.iea.org/data-and-statistics/data-product/world-energy-balances
15. IEA (2023d) World energy investment 2023. https://www.iea.org/reports/world-energy-investment-2023
16. Jamshidi M, Askarzadeh A (2019) Techno-economic analysis and size optimization of an off-grid hybrid photovoltaic, fuel cell, and diesel generator system. Sustain Cities Soc 44:310–320
17. Jong S, Sterkx S (2010) The 2009 Russian-Ukrainian gas dispute: lessons for European energy crisis management after Lisbon. Eur Foreign Aff Rev 15(4):511–538
18. Kim DW, Kim YM, Lee SE (2019) Development of an energy benchmarking database based on cost-effective energy performance indicators: a case study on public buildings in South Korea. Energy Build 191:104–116
19. Ozili PK (ed.) (2022) Global economic consequence of Russian invasion of Ukraine. In: Working Paper. Global economic consequence of Russian invasion of Ukraine. https://doi.org/10.2139/ssrn.4064770
20. Ozili PK, Ozen E (2023) Global energy crisis: impact on the global economy. In: The impact of climate change and sustainability standards on the insurance market. pp 439–454
21. Sweeney JL (1984) The response of energy demand to higher prices: what have we learned? Am Econ Rev 74(2):31–37
22. Sweeney JL (2002) The economics of energy. Edward Elgar Publishing, Cheltenham, UK
23. Zhang YJ, Bian XJ, Tan W, Song J (2017) The indirect energy consumption and CO_2 emission caused by household consumption in China: an analysis based on the input-output method. J Clean Prod 163:69–83

Chapter 3
Geopolitics of Oil and Trade War

Abstract Oil plays a significant role in global policy and geopolitics. As a scarce resource, it is central to political dynamics in the Middle East. During the Cold War era, the US and USSR influenced the Middle East because of their geopolitical influence on the oil market and the Suez Canal. This chapter analyzes the geopolitical importance of the Middle East and Venezuela in the context of the geopolitical importance of oil. After the post-Cold War period, Russian influence declined in the Middle East. This chapter examines the crucial role played by Saudi Arabia, Venezuela, Iraq, and the Gulf War; Libya, Yemen, Sudan, South Sudan, Nigeria, the Democratic Republic of the Congo, and the Republic of the Congo; Angola, Gabon, Equatorial Guinea, Cameroon, Gabon, Equatorial Guinea, and the Soviet Union; and Afghanistan, Ukraine, Russia, Great Britain, and the U.S. in the geopolitics of oil and trade politics.

Keywords Cold war · USSR · USA · Middle East · Crude oil · Energy · Trade war

3.1 Geopolitical Interest in Oil

In global politics, there are no permanent friendships or trade alliances; these relationships shift over time, driven by profit motives and the pursuit of state interests inherent in global trade. The geopolitical interests of states are important elements in global trade [19]. In the twenty-first century, oil, gas, and earth minerals are vital for all nations, especially Western China. This interest becomes evident in the trade with respect to geopolitics. The location of states in a region is crucial for their role in geopolitics, particularly in terms of trade and economic development. The supply of oil and earth minerals is crucial for all nations because of the scarcity of these resources and increased demand. Conflicts and civil wars in supply-side countries or in regions increase prices. The conflict approach is beneficial for controlling oil prices in the global market, as Western oil companies have the upper hand in their investments, particularly in oil-rich countries, which are highly invested in various oil manufacturers, with the support of Western countries. There would be a balance of trade strategies in the oil market among the GCC countries, Iran, and Western countries. They receive indirect benefits from conflicts, such as civil wars or proxy, which impact oil prices in the global oil market.

3.1.1 Venezuela

Venezuela is a major oil producer, and the majority of its oil exports go to the U.S. At the same time, they faced American sanctions, as did the American company Chevron, through joint ventures with PDVSA and other NATO countries' oil companies, which operate in Venezuela. Geopolitically, Venezuela is isolated from the oil market verbally, but the reality is that many Western countries are cooperating with the Venezuelan oil market (see strategic locations in Fig. 3.1). Historically, Venezuela faced problems with the United States over oil trading because of the nationalization of oil companies during the time of Hugo Chavez, which affected the geopolitical interests of the United States [12]. The nationalization of oil companies affected the interests of the United States in South America [37]. As a retaliatory measure, the U.S. introduced sanctions against Venezuela to counter Venezuela's oil production. This sanction affects the production of oil in Venezuela in the global market. Therefore, the oil supply of Venezuela does not affect oil pricing in the international market because of dollar sanctions. Iran's and Venezuela's oil pricing strategies are weakened in the global market because of American sanctions, oil prices can be used as weapons against those who are against the status quo power structure of the United States. The NATO countries have greater control over oil politics in Venezuela.

3.1.2 Iraq and the Gulf War

Saddam Hussein attacked Kuwait in 1991 to control the supply of oil for the high price of Iraqi oil in the global market [3]. During the invasion of Kuwait by Saddam Hussein, oil prices were high in the global market. Saddam Hussein had an economic motive for obtaining a higher price for oil. Iraq and Iran experienced a war during 1980–1988, and through this war [32], Iraq faced a severe economic crisis. This war is supported by Kuwait, Saudi Arabia, and other Arab countries [18]. After this war, these countries were not ready to decrease their production to increase the price. This is the motive for attacking Kuwait. These countries had already given various types of financial support to Saddam Hussein during the time of the Iran–Iraq War [52]. Nevertheless, Iraq went through a debt trap with Saudi Arabia and Kuwait after the war with Iran [40]. This impacted Iraq's economy, and these factors aggravated Kuwait's invasion. After the invasion of Kuwait, Iraq faced various sanctions from the UN for selling its oil until the U.S. invaded Iraq in 2003 [21]. After the invasion of the U.S. in Iraq, the U.S. controlled the oil. The U.S. took necessary action to repair the

3.1 Geopolitical Interest in Oil

Fig. 3.1 Strategic locations. *Source* Google

damage caused by Saddam Hussein in the Kuwait War. Kuwait and the United States continue to receive compensation from Iraq's oil revenues for the damage sustained during the Gulf War [7, 14]. The US achieved its geopolitical interest in oil in Iraq through the dismantling of the Saddam Hussein dictatorship through an invasion. After the Iraq invasion by the United States, it had the upper hand in controlling oil production in Iraq. The U.S. had indirect and direct control of the OPEC countries through its economic and military capabilities.

3.1.3 Libya

The United States supported rebels in 2011 to crack down on the Muammar Gaddafi dictatorship in Libya [8]. This enabled U.S. oil companies to control oil production in Libya through U.S. oil diplomacy [6]. During the Gaddafi era, the oil industry was nationalized. Therefore, the United States did not have adequate control over Libya's oil market. After this invasion, the United States controlled the Libyan oil market. This influenced the oil price in the global market, Saudi Arabia benefited from the Iraq and Libya invasions by the United States. The geopolitical interest of the U.S. in Libya is oil.

3.1.4 Yemen Conflicts

Yemen has been geopolitically significant since the Suez Canal opened [51]. The Gulf of Aden and the Red Sea are important for vessel transport in the global maritime economy. Oil is exported from Saudi Arabia to Southern Asia and Eastern Asia through the Gulf of Aden. The shipment from Asia to Europe involved the use of the Gulf of Aden. Yemen is a strategic location, especially during the Cold War—the major world powers were involved in Yemeni politics. During the Cold War era, Yemen was in two separate states: North Yemen and Communist South Yemen. North Yemen was supported by West and GCC countries, and South Yemen was supported by socialist blocs. Yemen's geopolitical proximity is important for global politics. After 1990, Yemen was unified with civil war. The Iranian Shia proxy Houthis constitute an important rebel group in Yemen. Houthis focused on Saudi interests in Yemen. Yemen has rich oil reserves [10], but these reserves have not been utilized because of their close relationship with GCC countries. Yemen has 3 billion oil reserves, and they are not utilizing its oil reserves because of civil war. Suppose Yemen had no civil war and good production capacity for oil. Yemen could have replaced many GCC countries that are dealing with Southeast Asia and South Asia due to the geopolitical advantage of Yemen. The interests of the US and GCC countries in Yemen are closely associated with the geopolitics of oil. Iran supported proxies from Yemen; Yemen-based forces had been attacking oil vessels in the Gulf of Eden [27]. This, in turn, affects the oil price because of the geopolitical importance of Yemen in the Middle East. Yemen has faced a lack of a stable government system since 1991. Conflicts in Yemen's coastal areas have led to price increases in oil. Thus, Iranian proxies attempt to stabilize the oil market and benefit from increasing the price of oil in the Middle East.

3.1.5 Sudan and South Sudan

Sudan and South Sudan together have an oil reserve of approximately 5 billion barrels [64, 65]. South Sudan has a more stable government, but it faces the challenge of landlocked geography. South Sudan uses Kenyan ports to export crude oil to different countries [31]. Sudan is very close to the Red Sea, but owing to civil war problems in Sudan, it has less production capacity in the global oil and gold market. Sudan is closely associated with GCC countries because of its religion and influence on the Arabic language. Saudi Arabia and the UAE are closely associated with civil war in Sudan through the export of Arms to Sudan. The Civil War in Sudan affected oil and gold production and export. This will indirectly support Saudi and the UAE oil and gold export interests in the global oil and gold market. Sudan is a significant country because of its geography and reserves of oil and gold. The geopolitical interest in oil and gold is clear in Sudan from the dominant countries of the GCC. Sudan has a critical geography for exporting oil through the Red Sea [44, 66]. The internal conflicts create problems in building stable production and export facilities in the region. Sudan and South Sudan are important countries in the context of the geopolitics of oil. Internal conflicts and civil wars are major problems for both countries in competing with the GCC countries in the oil market. Most oil and gold are exported to GCC countries from Sudan [61, 62]. Sudan faces serious sustainability development problems in its economy while having enough natural resources due to political and religious obedience with GCC-dominant countries.

3.1.6 Nigeria

Nigeria is a significant player in the oil market. The Islamic terrorist destabilizes the significance of Nigeria in the global oil market. Internal conflicts and terrorism reduce the development and growth of Nigeria. The NATO countries are investing in Nigeria. The Nigerian National Petroleum Corporation (NNPC) is the state oil corporation and is the largest oil company in Nigeria. The U.S. and U.K. oil companies play key roles in Nigeria.

3.1.7 Democratic Republic of the Congo and Republic of the Congo

Figure 3.2 shows the distribution of Crude Oil reserves in Central African Countries such as the Republic of Congo (38%), Gabon (27%), Chad (20%), and Equatorial Guinea (15%). The DRC and the Republic of the Congo are rich oil reserve countries in Central Africa. The Republic of the Congo has significant production capacity and export ability. The French company Total Energy is the largest oil-producing company in the Republic of Congo-Brazzaville. Perenco is a French multinational oil producer and the largest producer from the DRC [47]. Both countries have seaport facilities to export their oil to the global market. The civil war in the DRC is a serious problem for utilizing their opportunity in the global oil market. Congo-Brazzaville is also part of OPEC. The majority of offshore fields are controlled by Total Energy and its subsidiaries in Congo-Brazzaville. Both countries have significant potential in the growing oil market. To utilize this opportunity in the global market, a stable government without conflict and civil war is essential.

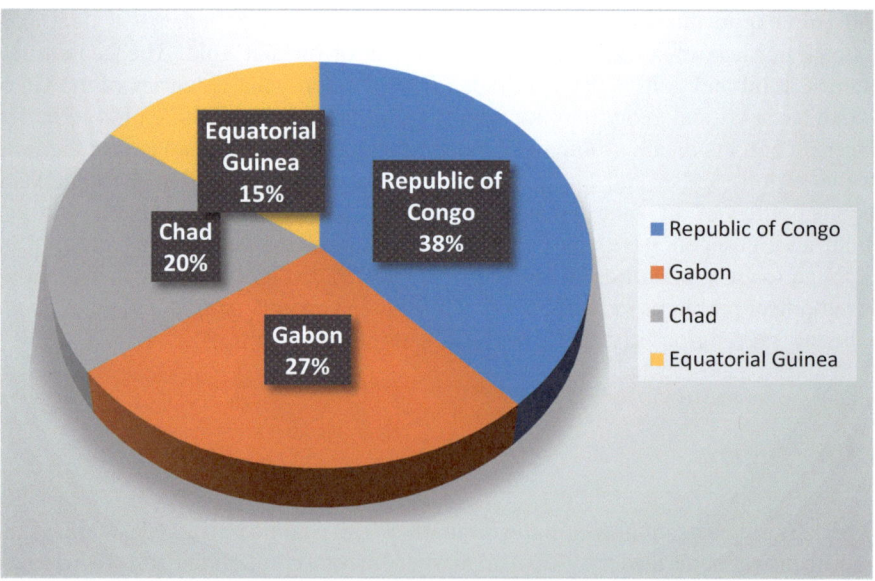

Fig. 3.2 Distribution of crude oil reserves in Central Asia as of 2021. *Source* Author's illustration using African Energy Commission, Statista 2024

3.1.8 Angola

Angola plays a crucial role in oil exports. Since 1991, international oil companies have played a vital role in Angola's oil production [11]. The grade of oil is suitable for Asian countries' refineries. In 2023, Angola exited OPEC due to disagreements about production quotas. Angola has a stable government system, which would be beneficial for increasing its global oil production and sales.

3.1.9 Gabon, Equatorial Guinea, and Cameroon

The cooperation of Central African countries can play a vital role in the geopolitical role of oil in global politics. The oil producers from Central Africa cooperate with each other for institutional cooperation in terms of oil production and supply. This has led to the overall development of Central African countries. A stable government system and a better education system are crucial for avoiding conflicts in Central African countries. The income from the oil should be invested in improvements in the education system for long-term success. Equatorial Guinea has 1.1 billion oil reserves. It is a very important oil producer in Central Africa; the oil industry is controlled by multinational giants from the USA and France. GEPetrol is the national oil company of Equatorial Guinea.

Cameroon has 2 billion oil reserves and significant oil production in Central Africa. NATO countries' oil companies are controlling the Cameroon oil market. The hydrocarbon corporation (SNH) is a major player in Cameroon under state ownership. It plays a major role in Cameroon's oil resources and partners with multinational companies from NATO countries.

3.1.10 The Central African CFA Franc Currency and the Oil Market

The France and NATO countries have significant influence in Central Africa through the XAF currency. The XAF currency was stabilized by the French treasury. The depreciation of the XAF currency is helpful for EU countries to buy oil from Central African countries. XAF currency is the neocolonial agenda of France to exploit the oil reserves and minerals of Central African countries. Six Central African countries use XAF currency: Cameroon, the Central African Republic, Chad, the Republic of the Congo, Equatorial Guinea, and Gabon. These countries are rich in oil and minerals. Since their independence, the six countries of the CAEMC have pegged their currency, the CFA franc (XAF), to the French franc. In 1999, following the launch of the Euro currency, the peg shifted from the French franc to the Euro currency [50].

3.2 Oil- and Mineral-Based Neoliberal Institutions for Central African Countries.

Multilateralism in Central Africa, which involves cooperation among all oil and mineral producers, will lead to a stable government system, such as the GCC. The GCC countries are not democratic, but they are theocratic monarchies under tenth-century Sharia law, and they are stable government systems. In Central Africa, all countries are democratically republics; however, internal conflicts hinder institutional cooperation. Central African countries, such as the GCC countries, should maintain cooperation. Civil war conflicts can decrease in Central Africa if the NATO countries wish to stop them. An embargo of arms to civil war groups will solve this issue. NATO countries are more interested in business, and their relative gains are greater than those of Central African countries. Institutional negotiations reduce the influence of Western countries in Central Africa to counter neocolonialism. Establishing venture partnerships will be a better option for Central African countries, such as Aramco in Saudi Arabia. Control over one's own resources is very important for the sovereignty of all nations.

3.2.1 Petrodollars and GCC Countries

Saudi Arabia is a strategic location in the GCC. The Saudi Kingdom is closely associated with the United States. Saudi Arabia is the largest oil producer within OPEC. The U.S. and Saudi Arabia were closely associated during the Cold War era. U.S. companies have invested in the oil companies of the GCC countries. Saudi Arabia is the leading power in the GCC countries. Saudi countries and the U.S. were closely associated with each other in Afghanistan to counter the liberal government of Afghanistan during 1979–1991. This alliance helps the US improve its relationship with Saudi Arabia and other GCC countries, particularly the US, to achieve geopolitical interests in oil through petrodollar agreements between Riyadh and Washington. The petrodollar agreement helped the US gain currency interest in global politics [54]. The confidence in the U.S. dollar has increased in global trade, and it has become a global reserve currency. Saudis also received tremendous support for Monarchy and Wahhabism in global politics from the U.S., which invested dollars in various portfolios in the U.S. Saudis received technical and military support from the U.S. against Iran's confrontation with the Sunni monarchy in Saudi Arabia. After 1979, Israel–Saudi conflicts were reduced with the support of the U.S. in Saudi Arabia.

Many GCC countries traded their oil for the dollar after the implementation of petrodollar agreements with Saudi Arabia and the US. Saudi Arabia plays a crucial role in GCC politics because of its geographical location in the GCC countries. The geopolitical influence of oil is very high for Saudi Arabia in global politics because of the large amount of oil reserves in the world and its production capacity. Saudi Arabia regulates oil prices by monitoring supply-side policies.

3.3 Oil as a Geopolitical Weapon

Oil is the lifeblood of every economy. Some countries worldwide do not produce enough oil. Owing to the high price of oil in the international market, oil-producing countries control their economic and military capabilities. During the war between the two nations, oil can be a weapon to protect one nation. During the Arab War, Saudi Arabia stopped supplying oil to the United States and Israel, which increased the price of oil in the international market. The U.S. did not have enough oil reserves to meet domestic oil needs. This led to stagnation and hit the U.S. economy hard. The oil embargo in the 1970s was a weapon against those who were not allies with oil producers against Israel. The Israeli war is closely associated with Palestinian liberation. During this war, those who were against Palestine had to deal with the embargo on oil [43]. This especially targeted the U.S. and Israel and affected the U.S. economy because of higher oil prices. The oil supply can be used as a weapon during times of conflict between nations. The war between Israel and Egypt's allies affected the oil price. Egypt is geopolitically significant because of its location and Suez Canal. The war affected the global shipment of oil through the Suez Canal, it also impacted the oil price. Conflicts in Red Sea countries affect oil pricing if they target vessel transport in the Red Sea. The conflict in Yemen affected oil transport vessels in the Gulf of Aden and the Red Sea. To counter the conflicts in Middle East America, the geopolitical interest of oil in Venezuela is very significant. Venezuela became more important again for NATO countries after the 2022 conflicts between Ukraine and Russia due to geopolitical interests in oil. If NATO countries wish to change their GCC oil diplomacy toward Venezuela, this would be a significant problem for GCC countries in the future. NATO countries can also control the oil market by changing the buying and selling of oil exports and imports.

3.3.1 Soviet Union Versus Afghanistan

After the Islamic revolution in Iran, the Soviet Union feared that Islamic radicals would address the situation on the borders of the Soviet Union in Central Asia. Afghanistan had a closer border with the Soviet Union. The growing fundamentalism in Afghanistan was a concern for Soviet Security after the Iranian revolution. Therefore, the Soviet Union military went to Afghanistan to support the Afghan government at the request of the Afghan government [28, 29, 57]. Eventually, this situation was utilized by Western countries to counter the influence of the Soviet Union in Afghanistan by funding Mujahideen rebels in Afghanistan. Moreover, the United States and Saudi Arabia used oil as a weapon against the Soviet Union by increasing the production of oil to counter the Soviet Union's oil exports. During the same years (1980–1988), Iraq and Iran experienced a war; during this time, oil prices decreased because Saudi Arabia increased production to counter the Soviet Union's oil export economy. The Soviet Union conflicted with Islamists in Afghanistan. As a result of this conflict, the Soviet Union needed more resources to secure the economy through the oil trade. However, this approach was not practical because of the lower price of the oil.

3.3.2 Ukraine and the Russian War

The Russian invasion of Ukraine severely impacted the global energy market [59, 60]. Russia invaded Ukraine in 2021 because of NATO's security concerns. After this invasion, the price of oil and natural gas increased due to sanctions against Russia. It affected the oil supply to Europe. Most EU countries are closely associated with Russia in terms of oil supply, which impacts the oil market. Sanctions against Russia impacted the Russian oil economy, and India came to support the purchase of oil from Russia at a price lower than the market price. This sanction did not affect Russian trade because of India's support. Moreover, the United States has attempted to increase its oil production in the market. The U.S. called for Saudi Arabia to increase the production of oil to counter Russian oil deals in the global market [26]. Saudi Arabia did not accept the United States' advice to increase oil production. This, in turn, favored Russia, the oil price has been on the rise since 2022. The sanctions against Russia in 2022 eventually impacted future transactions of the Petrodollar because of rising powers in the Global South.

3.3.3 America and Great Britain

During 1949–1970, American and British oil-producing companies such as Anglo-Persian Oil Company (BP), Gulf Oil (Chevron), Shell, Standard Oil of California (Chevron), Standard Oil of New Jersey (Exxon, later ExxonMobil), Standard Oil of New York (Mobil, later ExxonMobil) and Texaco (Chevron) controlled global oil production. It was important for the U.S. to gain control of the global oil market for exploration in the Middle East because of the technological capabilities of those companies. The Middle East and other developing countries did not have oil exploration technologies at that time. This eventually allowed the U.S. and Britain to invest in oil exploration. OPEC was established in the 1960s to counter the oil diplomacy of the United States in the Middle East. Eventually, OPEC countries established their own methods to control supply and production. This was against the U.S. interest, such as the oil embargo. The United States introduced the Petrodollar system with Saudi Arabia to counter the oil embargo in the future. This has been a positive approach to oil diplomacy in the Middle East since the Cold War by the U.S. [9].

3.3.4 NoN-OPEC

The U.S. has the world's most prominent man-made strategic petroleum reserve (SPR) [5]. This allows the U.S. to control the oil price globally. NOPEC is a bill passed by the U.S. Senate to counter the influence of OPEC and other producing nations engaging in the price and production of oil. This bill allows the U.S. government to sue OPEC members and other oil exporters under U.S. government law. The U.S. government has antitrust law under the Sherman Act for suing member countries against OPEC if they are against U.S. interests. This bill was passed by the United States Congress in 2007 by Senator Chuck Schumer, and it counters the dollarization of OPEC countries. This bill helped the United States counter geopolitical tensions in the Middle East. No OPEC bill had some domestic concerns among the United States oil producers. NOPEC legislation has led to tensions in the U.S. and oil-producing countries. However, this bill did not successfully achieve U.S. interests in the Middle East.

3.3.5 OPEC Behavior and Strategy

OPEC has followed various strategies to determine the prices and production levels of crude oil [2, 16, 24, 36, 48, 49, 58]. OPEC is a cartel of 13 oil-producing and exporting countries. The founding members of OPEC include Iran, Iraq, Kuwait, Saudi Arabia, and Venezuela. OPEC countries make up the majority of the world's oil production and regulate oil prices in the international market [22]. OPEC follows strategies to regulate the production of oil [1, 13, 16, 17, 23, 45]. During the energy crisis (pandemic), oil demand decreased considerably in 2020. Along with the price war between Russia and Saudi Arabia, which led to a dip in prices, OPEC further adjusted production to 7.2 million (b/d) in January 2021. The energy crisis worsened in 2022 during the Russia–Ukraine War. The war disrupted the supply of oil and natural gas from Russia to Europe; the demand–supply mismatch further increased energy prices.

OPEC was established to reduce the oligopoly power of U.S. and British companies in the Middle East. The strategies aimed to establish a favorable oil market per the member states' objectives. The sovereignty over natural resources in OPEC countries led to the oil crisis in the 1970s. The sovereignty of oil reserves was limited before OPEC. OPEC increased its influence on its energy resources and gained control over production. OPEC members are attempting to reduce market competition between member states. However, it was not practical because of the anarchic structure of international systems. OPEC accounts for 81.5% of the world's proven oil reserves and 38% of the total crude oil fraction [41]. As of 2025, OPEC has 13 members; in 2018, the Republic of Congo; in 2017, Equatorial Guinea joined OPEC; other members include Algeria, Angola, Gabon, Iran, Iraq, Kuwait, Libya, Nigeria,

Saudi Arabia, the United Arab Emirates, and Venezuela. Arab countries have more influence within OPEC [41].

3.3.6 OPEC +

OPEC is the original organization of oil-exporting countries, whereas the OPEC + border coalition includes non-OPEC member countries. This approach can help OPEC and non-OPEC countries effectively handle oil supply and pricing market dynamics. The OPEC + countries include Russia, Azerbaijan, Mexico, and Kazakhstan. OPEC + is coordinating with various non-OPEC countries for production cuts to stabilize the oil price. The Organization of Petroleum Exporting Countries Plus (OPEC +) member states, consisting of 12 OPEC members plus 11 other oil-producing countries, are the first to bear the brunt concerning energy demand security.

3.4 Energy and Trade War

Energy exports are a strategic instrument of the foreign policy of the U.S. The dominant powers in global politics have greater influence on trade negotiations. After the Second World War, America became a dominant power in trade. During the Cold War, the Soviet Union was a military power, but it was not a greater economic power than the United States. For this reason, the U.S. promoted various institutions in global politics, such as the IMF, World Bank, and GATT agreements (now WTO). The American trade negotiation structure included institutions and governments. American values of trade greatly influence these institutions, as the U.S. influence surpasses that of any other country in these institutions. These institutions promoted the values of the free market. The U.S. used economic interest as a tool against the Socialist bloc to achieve the United States' interests. The U.S. considered the energy trade a very important weapon for trade negotiations in global politics. The scarcity of conventional energy resources is a beneficial foreign policy for the U.S. through trade negotiations. After the Second World War, the United States achieved de facto integration with Western European countries through its Marshall Plan [46, 55, 56]. The success of the Marshall Plan increased Western Europe's productivity, which increased the various production capabilities of Western European countries. Western European countries supplied products to the U.S. The United States also supported the production of nuclear energy.

The U.S. and Western European countries cooperate in trade negotiations [4, 35]. Therefore, the U.S. sets low tariffs on Western European countries' exports to the U.S. This comparative advantage of producing various commodities in the U.S. is more expensive and less profitable than in Western Europe. Therefore, there is a trade-off. After the success of the Marshall Plan, the U.S. was interested in trade with Asian countries and adopted the same methods in Japan, South Korea, Thailand, and China during the Cold War period. In the Cold War era, those trade allies had higher exports to the United States. China introduced a new economic policy in the 1980s during the Cold War. This allowed Western investors to invest in China.

3.4.1 Trade Tactics After the Cold War

The historically low oil price era has reshuffled the system of alliances that was the cornerstone since 1973 [20, 42]. The U.S. learned lessons from the 1970s energy crisis due to high oil prices and stagflation, and the U.S. can no longer depend on other countries for its energy resources. Since the late 1990s, the U.S. has attempted to gather energy reserves from the rest of the world. After the Cold War, the influence of the Soviet bloc ended, and new countries were born in Europe and Central Asia. Russia and the other Soviet republics joined liberalization, globalization, and privatization campaigns through liberal economic institutionalism. This increased the demand for the dollar in the market. This allowed Western countries to invest in those nations, including India. This created ample opportunities and possibilities in those countries by creating trade dependency with the U.S. The U.S. used foreign direct investment as a tool to achieve its interests. China and India experienced substantial economic growth after 2000. As economic growth takes place, these countries require energy resources for their economy. China and India had comparatively fewer sources of crude oil. Therefore, they depend on the Middle East for oil. China has close cooperation with Brunei; China started its Belt and Road Initiative to establish cooperation with Brunei [25, 53]. During 1991–2000, until the Saddam Hussein era, India was closely associated with the Saddam Hussein Administration for Crude Oil [30]. Later, India depended on other GCC countries for oil. Following the Ukraine and Russia war in 2022, India strengthened its relationship with Russia for its energy demand, in turn, Russia supplied oil to India at a discounted price rather than the market price.

3.4.2 Effect of the U.S. Tariff on China

China conducts production for the entire world [15]. China is a member of the WTO trade agreement,[1] which allowed it to become a global exporter of consumer goods. China is one of the largest trading partners of the U.S. and the third-largest export market [38]. After 2010, domestic manufacturers in China adopted innovations and technologies to expand as did global manufacturers. This became a threat to the U.S. as the global position of American firms was challenged [33]. This significantly affected U.S. trade interests. Owing to the anarchy structure of international relations,

---

[1] https://www.wto.org/english/docs_e/legal_e/gatt47_e.htm

the U.S. wanted to sustain its trade hegemony. The U.S. used tariffs as a weapon against China to sustain its trade hegemony. This hurts the Chinese economy, as it affects its large-scale production. This creates a new power-balancing structure in global politics. This will reduce oil and energy consumption in China, as factories reduce production. This also has an impact on the oil market, there is less demand for oil by China, as they reduce their extent of production. To counter these trade tactics, the U.S. will look for the next best alternative, such as India or a U.S. ally. The U.S. tariff will allow its free market economy to flourish. The U.S is considered as a free-market economy and tariffs are instruments to attain global leadership in trade. This will, in turn, affect the global energy market as global production shifts.

A tariff is a tax imposed by a domestic government on imported commodities. Setting a tariff increases the price of goods and makes them less competitive than domestic goods. Tariffs are instrumental in fostering domestic trade. However, the implementation of tariffs has led to a trade war. An increase in the tariff on Chinese exports to the U.S. will lead to less demand for manufacturing products from China because of the high price of Chinese products in the U.S. market. As a result of the mutual trade war, China tends to lose more than the U.S. does [34, 39]. This ultimately decreases the manufacturing capacity of Chinese firms, significantly reducing energy consumption in China.

Figure 3.3 illustrates how the U.S. tariff impacts Chinese production. With an immediate effect of the change in price from P0 to P1, the quantity demanded and exported is reduced to Q0 (60) from Q1 (40). This results in a decrease in Chinese production, which tends to decrease production and exports to the U.S. The U.S. has exported to China, and an increase in tariffs on China will impact the U.S. economy. The after-effects of the tariff will protect American interests more than Chinese interests because of the United States dollar and energy diplomacy.

To understand the effect of tariffs on the energy market, an example of the U.S. import of solar panels from China will provide insights. If the U.S. imposes a tariff on imported solar panels, the price increases from P_0 to P_1 and reduces the quantity

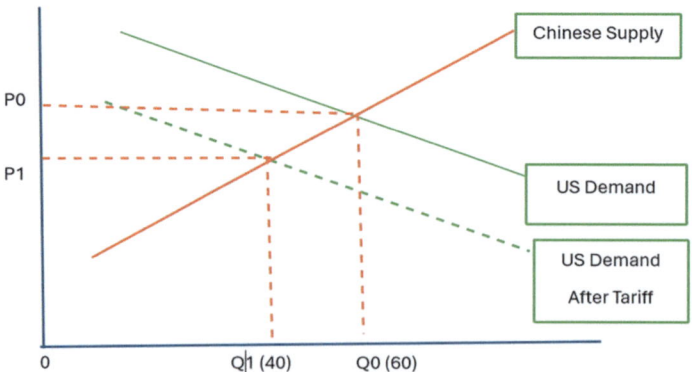

Fig. 3.3 Effects of tariffs on Chinese production Quantity of Chinese goods exported to the U.S. (x-axis) and price (y-axis)

of energy technology (solar panels) from Q_0 to Q_1. This highlights the trade-off: protecting the domestic industry versus slowing clean energy adoption due to higher costs. The increase in the price from P0 to P1, hence leads to the change in the quantity demanded (of solar panels) from Q0 to Q1. As a result of the tariff, the price remains high, the quantity of solar panels available decreases, and imports fall.

Less production in China will lead to less energy consumption, a scenario similar to that of the COVID-19 pandemic, as industrial activity was low, production fell, and the energy demand remained low. During the 2007–2008 recession and the COVID-19 crisis, the oil price fell; this pattern might have been repeated in the U.S. because of its current trade strategies. In the short run, there is no alternative to counter imports from China in the U.S. This leads to inflation in the economy, and people have to spend more money on their basic needs. Hence, there would be less energy consumption in the U.S. This would form a vicious cycle of low demand and low consumption of energy. This impacts energy prices, particularly those of crude oil. The price of crude oil decreases in the short run. This will impact global production and the global energy market.

Summary

Geopolitical interest in oil is significant for global energy security. The dominant countries attempt to achieve their self-interest in various oil-rich countries. The interest of superpowers always protects their national interest in oil-rich countries. Within the GCC, dominant countries tend to exhibit relatively greater fairness in terms of governance and economic transactions than Central Africa and Venezuela do. GCC countries have stable governments, often theocratic monarchies with fewer internal conflicts, and civil wars. The Western liberal democratic countries and the GCC maintain warm relations, although they have not followed democracy. African democratic countries face civil war and internal conflicts, and some countries are closely associated with Western liberal democracy. Owing to civil war and internal conflicts, they are not able to grow while they have enough oil reserves and minerals. Western democratic countries are closely associated with the GCC countries despite the latter not being democratic. Many countries in Africa face exploitation of their oil reserves and minerals by Western multinational corporations, while many oil-rich countries are engaged in internal conflicts or civil wars. This civil war and internal conflict did not significantly affect the interests of Western countries' oil investment, but at the same time, they achieved their self-interest from their investment. The economic and military interests of dominant powers create the growth of nations through trade and investment. Religion, language, and ethnicity are tools used by dominant countries to achieve geopolitical interest in oil and trade. Tariffs play a vital role in international trade. The tariff war will affect the trade volume of countries. The tariff plays a crucial role in oil pricing because a higher tariff reduces oil consumption. The demand for oil decreases, and the price of oil decreases. In the beginning of the twenty-first century, China stands as a world production house. The tariff war between China and the U.S. negatively affects China. Higher tariffs imposed by the U.S. towards Chinese products leads to a crash in the Chinese economy, and the oil demand in the short term until a new production house shifted.

References

1. Alhajji AF, Huettner D (2000) OPEC and other commodity cartels: a comparison. Energy Policy 28(15):1151–1164. https://doi.org/10.1016/S0301-4215(00)00089-2
2. Alhajji AF, Huettner D (2000) OPEC's production decisions: an empirical test of alternative hypotheses. Energy J 21(4):125–144. https://doi.org/10.5547/ISSN0195-6574-EJ-Vol21-No4-6
3. Alnasrawi A (1992) Iraq: economic consequences of the 1991 Gulf War and future outlook. Third World Quart 13(2):335–352
4. Aydin U (2015) Theoretical framework of the European union's energy security challenges in 21st century. Euras Stud J 2(2):48–60
5. Beaubouef BA (2007) The strategic petroleum reserve: US energy security and oil politics, 1975–2005, vol 21. Texas A&M University Press
6. Bini E (2019) From colony to oil producer: US oil companies and the reshaping of labor relations in Libya during the Cold War. Labor Hist 60(1):44–56
7. Blanchard CM (2010) Iraq: oil and gas legislation, revenue sharing, and US policy
8. Boyle FA (2013) Destroying Libya and world order: the three-decade US campaign to terminate the Qaddafi revolution. SCB Distributors
9. Clark WR (2005) Petrodollar warfare. Oil, Iraq and the future of the dollar
10. Colton NA (2010) Yemen: a collapsed economy. Middle East J 64(3):410–426
11. Corkin L (2017) After the boom: Angola's recurring oil challenges in a new context (No. 72). OIES Paper, WPM
12. Corrales J, Penfold-Becerra M (2011) Dragon in the tropics: Hugo Chávez and the political economy of revolution in Venezuela. Brookings Institution Press
13. Dahl C, Yücel M (1991) Testing alternative hypotheses of oil producer behavior. Energy J 12(4):117–138
14. Dickson M (2013) Legal issues in the united nations compensation commission on Iraq. JL Pol'y Global 14:21
15. Eichengreen B, Tong H (2006) How China is reorganizing the world economy. Asian Econ Policy Rev 1(1):73–97
16. Fattouh B (2007) OPEC pricing power: the need for a new perspective. Oxford Institute for Energy Studies
17. Fattouh B, Mahadeva L (2013) OPEC: what difference has it made? Annu Rev Resour Econ 5(1):427–443
18. Freedman L, Karsh E (1993) The Gulf conflict, 1990–1991: diplomacy and war in the new world order. Princeton University Press
19. Sweeney JL (2016) Geopolitical factors. Econ Energy 4(9):48. Department of Management Science and Engineering, Terman Engineering Center
20. Gordon DF, Reddy DP, Rosenberg E, Bhatiya N, Saravalle E (2018) US geopolitical challenges and opportunities in the era of lower oil prices. Center for a New American Security
21. Gordon J (2010) Invisible war: the United States and the Iraq sanctions. Harvard University Press
22. Greene DL (1991) A note on OPEC market power and oil prices. Energy Econ 13(2):123–129. https://doi.org/10.1016/0140-9883(91)90034-T
23. Griffin JM (1985) OPEC behavior: a test of alternative hypotheses. Am Econ Rev 75(5):954–963. https://www.jstor.org/stable/1821363
24. Griffin JM, Teece DJ (eds) (1982) OPEC behavior and world oil prices. Allen & Unwin, London
25. Hadi ARA, Hussain HI, Zainudin Z, Rehan R (2019) Crude oil price and exchange rates—the case of Malaysia and Brunei. Int J Finan Res 10(5):1–10
26. Heibach J (2024) The benefits of neutrality: Saudi foreign policy in the wake of the Ukraine war. Global Pol 15(4):789–793
27. Hokayem E, Roberts DB (2023) The war in Yemen. Survival 58(6):157–186. Routledge
28. Hughes G (2008) The Soviet-Afghan War, 1978–1989: an overview. Def Stud 8(3):326–350

References

29. Kakar MH (1995) The Soviet invasion and the Afghan response, 1979–1982. Berkeley, Los
30. Khan N (2010) IRAQ: a study of changing political dynamics since 2001. Doctoral dissertation, Aligarh Muslim University Aligarh (INDIA)
31. Kibanga LW (2017) The impact of trade ties on the economic relationship between Kenya and South Sudan. Doctoral dissertation, University of Nairobi
32. Klare MT (2013) Arms transfers to Iran and Iraq during the Iran–Iraq war of 1980–88 and the origins of the Gulf War. In: The Gulf War of 1991 reconsidered. Routledge, pp 3–24
33. Li C, He C, Lin C (2018) Economic impacts of the possible China–US trade war. Emerg Mark Financ Trade 54(7):1557–1577
34. Li C, He C, Lin C (2020) Economic impacts of the US–China trade war: evidence from stock markets. The World Economy 43(8):2165–2181. https://doi.org/10.1111/twec.12978
35. Luft G, Korin A (eds) (2009) Energy security challenges for the 21st century: a reference handbook. Bloomsbury Publishing, USA
36. MacAvoy PW (1982) Crude oil prices as determined by OPEC and market fundamentals. Ballinger Publishing Company
37. Monaldi F (2018) The collapse of the Venezuelan oil sector and its global consequences. Atlantic Council
38. Morrison WM (2017) China–U.S. trade issues. In: Congressional research service report, 7–5700. Congressional Research Service, Washington DC
39. Ngai P (2005) Global production, company codes of conduct, and labor conditions in China: a case study of two factories. China J 54:101–113
40. Nonneman G (2004) The Gulf states and the Iran–Iraq war: pattern shifts and continuities. Iran, Iraq, and the Legacies of War, pp 167–192
41. Organization of the Petroleum Exporting Countries (OPEC) (2023) Annual Statistical Bulletin 2023. https://www.opec.org/opec_web/en/publications/202.htm
42. O'Sullivan ML (2017) Windfall: How the new energy abundance upends global politics and strengthens America's power. Simon and Schuster
43. Painter DS (2014) Oil and geopolitics: the oil crises of the 1970s and the cold war. Histor Soc Res/Historische Sozialforschung 186–208
44. Patey L (2014) The new kings of crude: China, India, and the global struggle for oil in Sudan and South Sudan. Hurst
45. Percebois J (1989) Économie de l'énergie. Éditions Economica, Paris
46. Percebois J (1989) The oil producing countries and the world energy market: a review of the oil market in the 1980s. Kluwer Academic Publishers
47. Petrovic S, Zoucha J (2023) Congo. World energy handbook. Springer International Publishing, Cham, pp 37–52
48. Ramcharran H (2001) OPEC's production under fluctuating oil prices: Further test of the target revenue theory. Energy Econ 23(6):667–681. https://doi.org/10.1016/S0140-9883(01)00083-6
49. Salant SW (1976) Exhaustible resources and industrial structure: a Nash-Cournot approach to the world oil market. J Polit Econ 84(5):1079–1093
50. Samba MC (2018) Exchange market pressure in the Central African Economic and Monetary Community (CAEMC) Area: empirical assessment of the macroeconomic determinants. Int Econ J 32(3):470–482
51. Schmitz C (2001) National borders, the global economy and geopolitics: a view from Yemen. Geopolitics 6(2):79–98
52. Segal D (1988) The Iran-Iraq war: a military analysis. Foreign Aff 66(5):946–963
53. Slesman L, Bahar R Oil and gas dependence of brunei economy: current progress and challenges. In: Stability, growth and sustainability: catalysts for socioeconomic development in Brunei Darussalam
54. Spiro DE (1999) The hidden hand of American hegemony: petrodollar recycling and international markets. Cornell University Press
55. Tarnoff C (2018) The Marshall plan: design, accomplishments, and significance, vol 18. Congressional Research Service, Washington, DC

56. Trebat NM (2018) The United States, Britain and the Marshall plan: oil and finance in the early postwar era. Econ Soc 27:355–373
57. Westad OA (2005) The global cold war: third world interventions and the making of our times. Cambridge University Press
58. Ye R, Yang X, Zhou Y, Lin C, Chen Y, Chen J, Bian M (2025) Energy demand security in OPEC+ countries: a revised 4As framework beyond supply security. Energy 320:135261
59. Zhang Q, Hu Y, Jiao J, Wang S (2024) The impact of Russia-Ukraine war on crude oil prices: an EMC framework. Human Soc Sci Commun 11(1):1–12
60. Zhang X, Yu L, Wang S, Lai KK (2009) Estimating the impact of extreme events on crude oil price: an EMD-based event analysis method. Energy Economics 31:768–778
61. https://oec.world/en/profile/bilateral-product/crude-petroleum/reporter/ven
62. https://oec.world/en/profile/country/sdn
63. https://energycapitalpower.com/dynamic-oil-companies-operating-in-the-congo/?
64. https://www.worldometers.info/oil/yemen-oil/
65. https://www.worldometers.info/oil/sudan-oil/
66. https://theconversation.com/middle-eastern-monarchies-in-sudans-war-whats-driving-their-interests-251825

Chapter 4
Recent Trends in Energy Production and the Changing Energy Mix

Abstract This chapter examines traditional and non-traditional energy production trends in the UK, Poland, and India. These countries are in various stages of energy transition. The UK is a developed country committed to achieving net-zero emissions by 2050 with a strong focus on the transition to renewable energy. Poland's economy is in transition; it was once a major coal consumer, but is now shifting to renewable energy. Owing to its large population, rising consumer demand, and industrial growth, India still has a long way to go in transitioning to renewable energy. On the basis of the recent budget, India plans to sign agreements with US firms to adopt renewable energy from nuclear energy. Fossil fuels are nonrenewable resources; it takes long years to replenish oil, coal, and natural gas. Resource extraction should be optimized to maintain reserves in the future. This chapter examines the applicability of the Hotelling rule in the extraction of resources, such as those in the mining industry. The energy trends of the three selected countries during 2000–2022 are further discussed. The trend indicates that UK, is progressing rapidly in their transition and are investing heavily in the adoption of renewable energy production. Transition economies, such as Poland in Europe, still rely on pipelines from Russia for natural gas, as well as other energy commodities, to meet their industrial and residential energy needs. India may greatly benefit from the adoption of renewable energy. The externalities of production have a significant detrimental effect on the environment and air quality in India.

Keywords Hotelling rule · Resource extraction · Hotelling rule · Energy production trends

When a miner extracts fossil fuel resources on the basis of Hotelling's rule, the marginal net benefit of each period is required to rise at the interest rate, but this requires a decrease in the marginal cost of extraction for each period as per Hotelling Rule [1, 2]. For this purpose, the miner will need to reduce the amount of extraction for each period because the marginal cost increases as the extraction level increases. Thus, to meet the conditions of Hotelling's rule, the miner will extract fewer resources for each period and fewer resources in the next period than in the previous period.

Under Hotelling's rule, the miner should take the optimal mining path to extract fewer resources for the next period (t + 1) than the previous one (t).

Should oil companies limit their current oil extraction to maximize their future profits according to the Hotelling rule? Nonrenewable resources take a longer time to replenish. However, they can be extracted quickly and exhausted quickly. If mining companies follow the Hotelling rule, the extraction rate should be lower, keeping resources for the future. However, the market is volatile, and geopolitical factors are pivotal in energy price dynamics. Global energy demand continued to recover from the COVID-19 pandemic in 2022, and supply chain issues and conflicts in Ukraine continue to impact the international energy sector. Considering the complex dynamics, mining companies may not stick to the rule of extraction as the rule says; hence, there is less tendency to follow the optimum extraction rule and more volatile energy prices globally. Green energy transition reduces the demand for fossil fuels, and this will help in optimum extraction of fossil fuels.

In this chapter, we evaluate the trends in the Production of Conventional and Nonconventional Resources. Here, we analyze the trends in energy production in three countries: the UK, Poland, and India. We analyze the trends in coal, peat, oil shale, coal, natural gas, and renewables. The UK set an ambitious target of an 80% transition to renewable energy to achieve net-zero emissions by 2050. The UK is ahead of renewable technology adoption, and the data show that in 2022, the UK will account for 40% of electricity generation. For comparison, these countries were selected:

1. UK—a developed country with the fastest renewable energy transition.
2. Poland—transition economy from coal dependency.
3. India—a developing economy with high demand for fuel.

4.1 UK's Trends in Energy Production from 2000 to 2022

Figure 4.1 depicts the trends in the UK's coal, peat, and oil shale energy production; imports; exports; and total energy supply for the period 2000–2022. The total energy supply declined rapidly after 2014. The trends show the UK's transition toward renewables and its decarbonization strategies to reduce carbon emissions. The year 2015 marked a deliberate effort to phase out coal. The UK has adopted renewable energy in response to climate change. However, its market-oriented approach to renewable energy was slower than that of other EU countries during this period [3]. The UK deployed renewable energy technologies to achieve a target of 20% of its energy from renewable sources by the year 2020 [4]. The trend shows that the UK will replace coal with investments in renewables.

The trend in crude oil, natural gas liquids, and feedstock production during 2000–2022 shows a decline in the total energy supply (see Fig. 4.2). Domestic production declined during 2000–2014 as the North Sea's gas reserves were depleted. However, LNG and natural gas are imported via pipelines from Norway, Belgium, and the

4.1 UK's Trends in Energy Production from 2000 to 2022 41

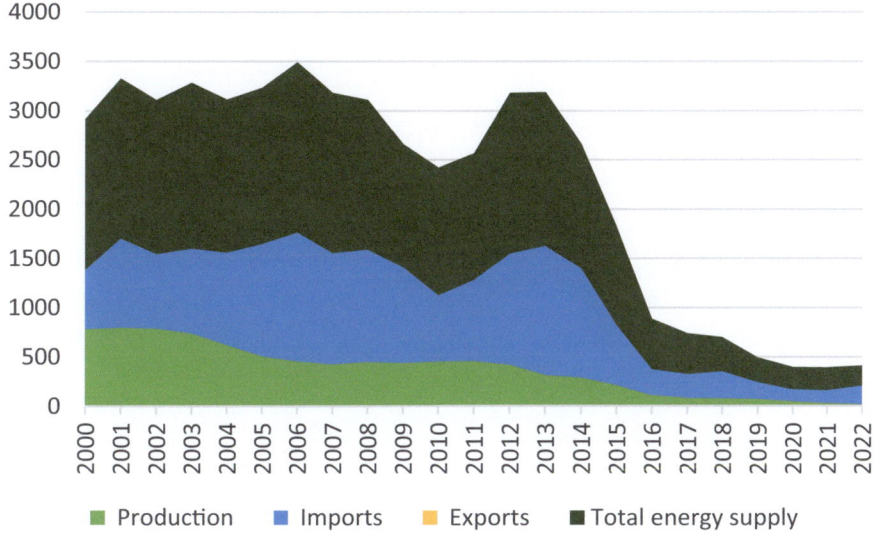

Fig. 4.1 Coal, peat, and oil shale in the UK during 2000–2022

Netherlands. The trend shows a declining trend in exports and an increase in imports in the UK.

The UK's oil production decreased drastically from 2005 to 2014. In 2015, oil production decreased to 70% of its original level in the 2000s. Domestic oil production was reduced, and oil imports increased in 2012. In 2022, imports constituted

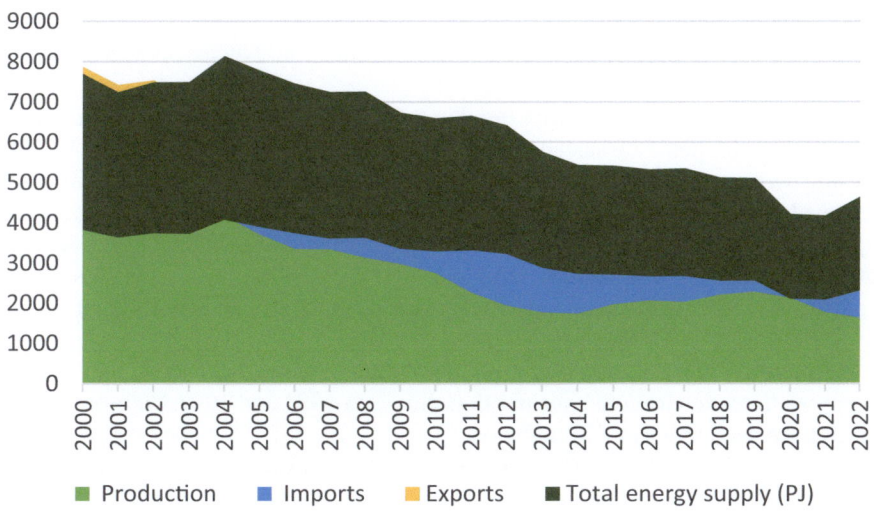

Fig. 4.2 Crude, NGL, and feedstocks in the UK during 2000–2022

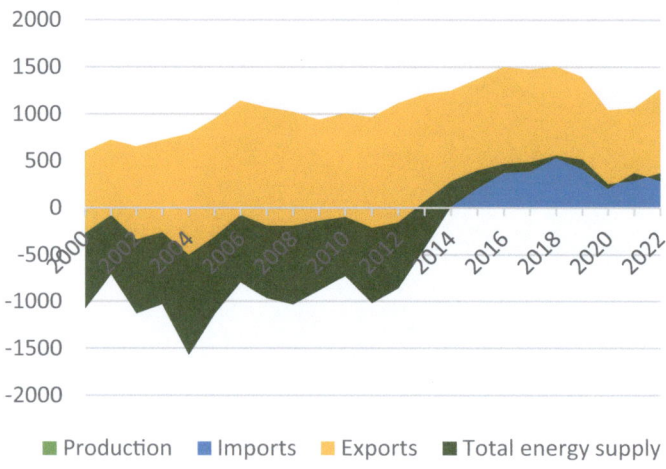

Fig. 4.3 Oil, UK

50% of the total energy supply, (see Fig. 4.3). At present, the UK oil refining industry faces socio-political pressures to decarbonize its production. Low-carbon reorientation is possible in declining industries as long as it is profitable [5]. The trend showed that the UK's reliance on imports aligned with its sustainability goals.

The UK's natural gas supply dynamics are illustrated with the production trends during 2000–2022 in Fig. 4.4. Domestic production accounted for 65%, whereas imports increased to 50% of the total natural gas supply in the UK. Britain's vast natural gas imported via pipelines from the Netherlands, Norway, and Belgium is meant to meet the UK's energy goals. Natural gas has a vulnerable carbon intensity, as do crude oil and coal [6].

The UK tries to decline its nuclear energy production. Nuclear power is part of the UK's energy security (see Fig. 4.5).

Figure 4.6 shows the increasing trend in renewable energy generation from 2010. The UK has the goals of sustainability and achieving net-zero targets [7, 8]. The UK utilizes its domestic resources efficiently, thereby increasing solar adoption and the development of onshore and offshore wind farms. The UK's production and supply of renewable energy gained momentum during 2000–2022. This trend indicates successful decarbonization strategies and a transition away from fossil fuel production. The UK utilizes wind, wave, tidal, solar, hydro, and geothermal energy [9].

Historically, the UK has relied on fossil fuels. Since the 1960s, the UK has attempted to reduce its coal consumption. Between 1984 and 1997, 70% of the coal mines were decommissioned. In the past few decades, the UK has extracted natural gas from the North China Sea. However, the UK imports natural gas via pipelines from Belgium, Norway, and the Netherlands. The UK has undergone a transition to renewable energy in the recent past and is committed to achieving net-zero emissions by 2050. The UK government has declared the need to reduce carbon emissions by

4.1 UK's Trends in Energy Production from 2000 to 2022

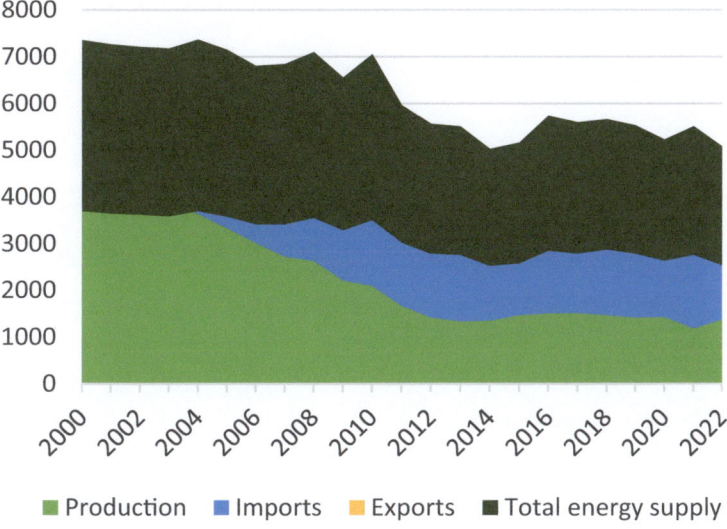

Fig. 4.4 Natural gas, UK

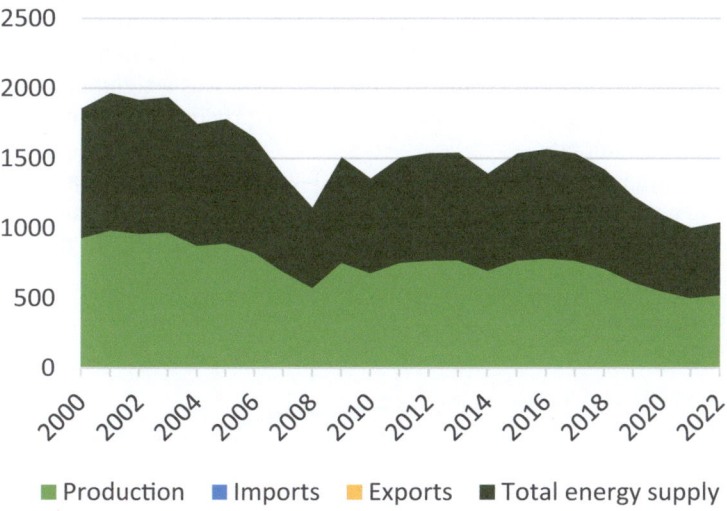

Fig. 4.5 Nuclear energy in the UK

2035 to achieve the net-zero target. Cities and councils across the UK have made their commitment to achieving this goal. Compared with other countries, the UK has come far from reducing emissions. The government has set ambitious targets through its 10-point green plan by building 40 GW of offshore wind by 2030, switching to EVs, and ending the use of petrol and diesel cars by 2030.

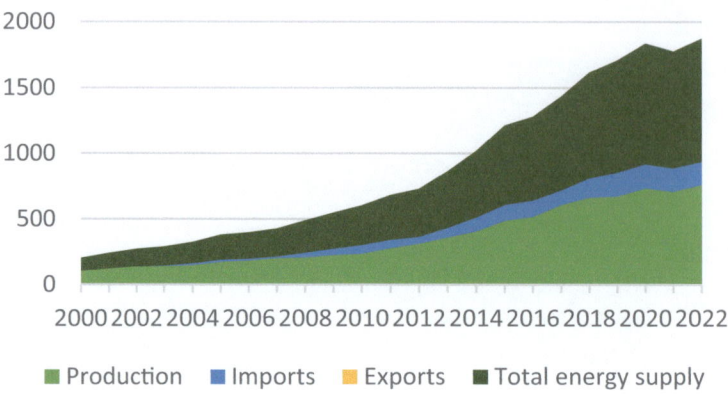

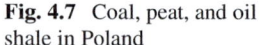

Fig. 4.6 UK's trends in renewables and waste

4.2 Poland's Trends in Energy Production from 2000 to 2022

Poland has been reducing coal, peat, and oil shale production for years. There has been a significant decline in output across the years, and production has declined by 2022. The total energy supply is stable and shows a negative trend. This is due to the dependence on imports. From 2016 to 2020, there was a significant increase in imports of coal (see Fig. 4..4.7). Amid the decline in production, Poland is balancing the shortage with imports.

Poland oil imports were stable during 2000–2014, and there was a sharp decline in 2015 and a surge in 2016. The increase in imports reflects the reliance on foreign markets. The total energy supply decreased during 2015–2016 and then increased

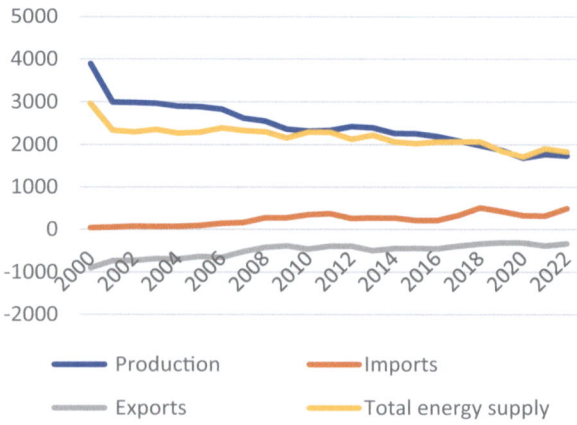

Fig. 4.7 Coal, peat, and oil shale in Poland

4.2 Poland's Trends in Energy Production from 2000 to 2022

after 2016. The significant cuts during 2014–2016 were due to refinery shutdowns and market fluctuations amid global oil market dynamics and geopolitics (Fig. 4.8).

There was low growth in the natural gas production of Poland from 2000 to 2022. Imports are the primary source of natural gas in Fig. 4.9. Russia is a leading supplier of natural gas. The demand for natural gas for industrial and domestic purposes is increasing. Reliance on foreign suppliers is due to national energy security concerns. The fossil fuel price slowdown in 2020–2021 was followed by an increase in energy prices in 2022 as a result of the supply impact from the Russia–Ukraine war.

The trends in renewable energy production in Poland are shown in Fig. 4.10. Self-sufficiency goals necessitated the transition to renewable energy, and exports and imports were minimal during this period. Poland's renewable energy mix includes

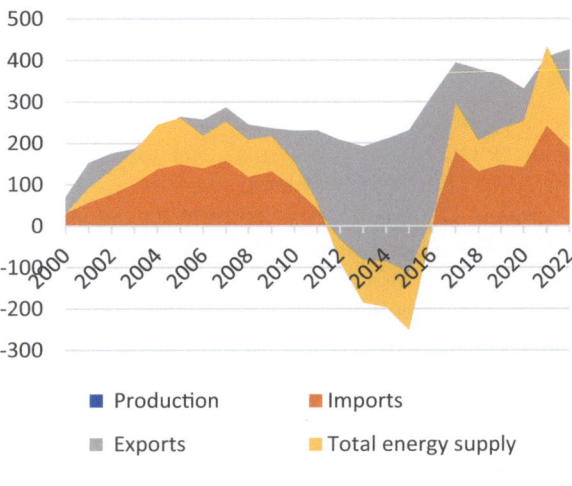

Fig. 4.8 Oil products in Poland

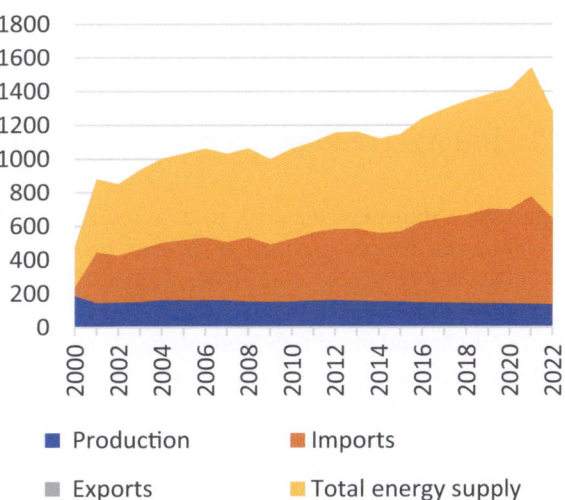

Fig. 4.9 Natural gas in Poland

Fig. 4.10 Renewables in Poland

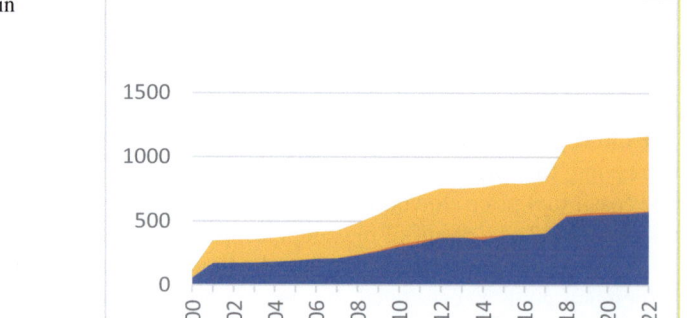

solar, wind, hydro, geothermal, and biomass. Investments in renewable energy increased after 2015 in Poland.

Poland is geographically close to Russia and Germany. Poland has anticipated threats from Russia since 1991. Poland is an EU and NATO member, but, at the same time, Poland is also anticipating threats from Germany too. Poland imported natural gas from Russia until April 2022. After Russian military aggression in Ukraine, Poland stopped buying gas from Russia, which impacted the Polish economy with high inflation. Poland started importing gas from the Baltic Pipe from Norway. This led to a higher price of gas. Poland imports crude oil from Saudi Arabia, Norway, the United States, Nigeria, and the UK. Poland is a transition economy; the higher price of crude oil and LPG led to an economic crisis in the short run. This impact can lead to stagflation. High prices of crude oil and natural gas lead to domestic political confrontation.

4.3 India's Trend in Energy Production from 2000 to 2022

Figure 4.11 illustrates the coal, peat, and oil shale trends during 2000–2021. India has domestic energy production to meet growing needs. Seventy-five percent of the total energy supply comes from domestic production, and 25% is imported. Excessive demand from the ever-increasing GDP and population highlights India's heavy reliance on fossil fuel production.

Figure 4.12 shows the trends in crude, NGL, and feedstocks in India during 2000–2021. The energy supply increased in 2017. After 2017, the total energy supply declined, indicating a reduction in imports. Nearly 78% of the energy supply comes from imports, indicating a dependence on imports. In 2021, the energy supply was

4.3 India's Trend in Energy Production from 2000 to 2022

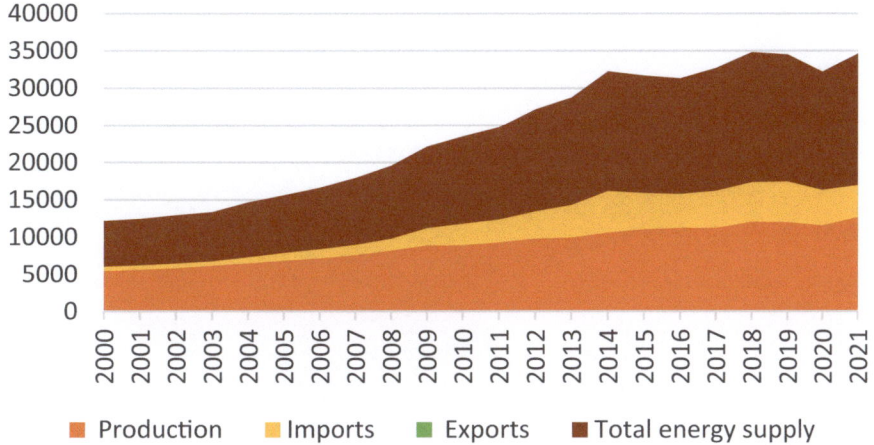

Fig. 4.11 Coal, peat, and oil shale in India

22,500 units, with the production of only 5000 units and the central part of the energy as imports with 17,500 units.

India is an importer of oil. However, India was a major oil refinery and a net exporter of byproducts of crude oil during this period. The contribution of oil to the total energy supply is harmful for green energy transition. India imports crude oil for the refineries. By products exports increased significantly during the mid-2000s, spiked in 2012, and then stabilized, with some fluctuations afterward (see Fig. 4.13).

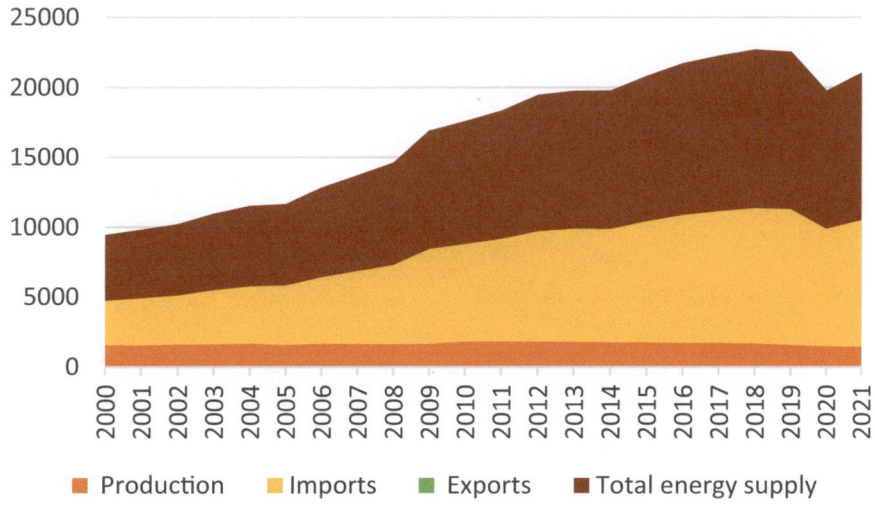

Fig. 4.12 Crude, NGL, and feedstocks in India

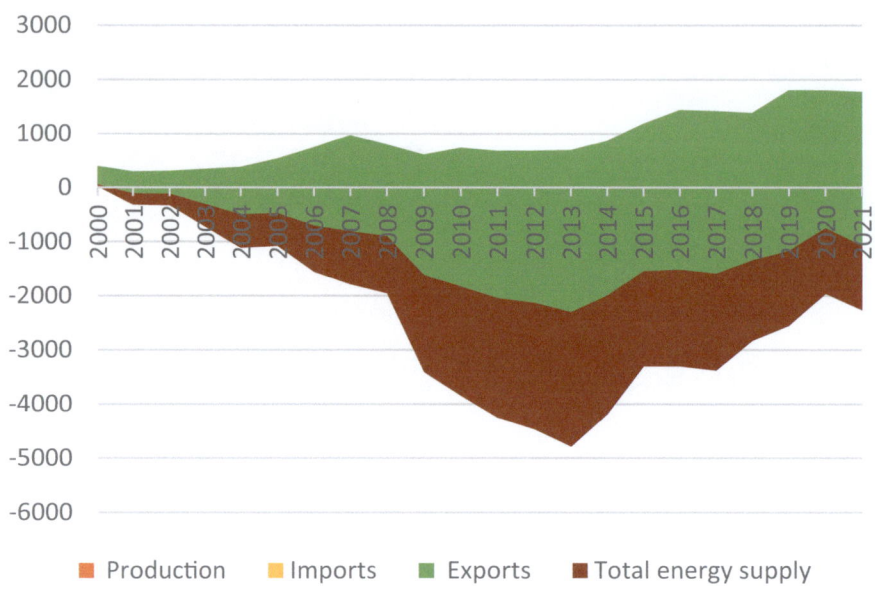

Fig. 4.13 Oil products in India

Figure 4.14 shows the natural gas production, imports, exports, and total energy supply during 2000–2021. There has been stable growth since 2000, but the total energy supply has steadily declined since 2011. Domestic production spiked in 2010 and then declined. The decrease in output increased. There is a dire need to invest in domestic production and alternative renewable energy channels to foster energy security.

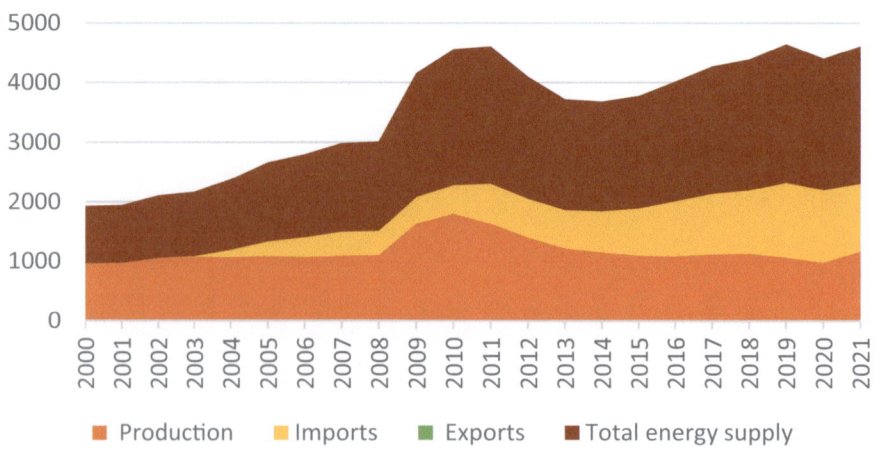

Fig. 4.14 Natural gas

4.3 India's Trend in Energy Production from 2000 to 2022

India is an emerging economy in South Asia. India experienced serious problems during the time of the Gulf War because of the high price of crude oil. India approached this situation with a new economic policy for maintaining an adequate foreign reserve for buying oil from the Middle East. The collapse of the Soviet Union led to serious problems in the Indian economy until the adoption of new economic reforms. India had an oil trade with Saddam Hussain. In 1995, the United Nations adopted the Oil-for-Food Programme in Iraq [10]. India had ties with Iraq for the Oil-for-Food Programme until 2003 [11]. It was also beneficial for India to address higher oil prices. India has limited resources of natural gas and petroleum, whereas India has many refineries for the extraction of crude oil. India has various types of trade cooperation for fossil fuel with Qatar, the UAE, and Saudi Arabia. India has a hedging approach for buying crude oil at a lower price. India imports LPG from Qatar, Saudi Arabia, and the UAE. Saudi oil companies are investing in India to establish new refineries. India and Russia also maintain a good relation historically from Cold War to mulipolar world order. India's and Russia's oil trade increased after 2022, with Russian company Rosneft collaborating with Nayara Energy company in India since 2017. India exports LPG and petroleum products in various countries. Most of the South Asian countries depend on the Indian fossil fuel industry.

Summary

The conventional and nonconventional energy production trends in India, Poland, and the United Kingdom are examined in this chapter. These nations' energy transitions are in various stages. With a strong emphasis on the transition to renewable energy, the developed nation of the United Kingdom is dedicated to achieving net-zero emissions by 2050. The economy of Poland is changing; it used a large amount of coal, but now, it uses more renewable energy. India still has a long way to go in its transition to renewable energy because of its large population, growing consumer demands, and industrial development. According to the latest budget, India intends to enter into contracts with US companies to switch from nuclear to renewable energy. Nonrenewable resources such as fossil fuels take a long time to replenish. Nuclear energy is not an alternative for India to reduce carbon emissions due to its higher population. India needs to attract green energy FDI to establish new joint ventures with different stakeholders to improve the green energy transition.

Questions

1. Explain how oil companies use the Hotelling rule to limit their current oil extraction rate to maximize their future profits.
2. Discuss how the energy markets driven by fossil fuels are being replaced by renewable energy sources.
3. What are the geopolitical implications of the U.S. as the world's largest oil producer?

References

1. Dasgupta PS, Heal GM (1979) Economic theory and exhaustible resources. Cambridge University Press
2. Hotelling H (1931) The economics of exhaustible resources. J Polit Econ 39(2):137–175
3. Elliott D (2016) Renewable energy in the UK: a slow transition. Green economy reader: lectures in ecological economics and sustainability. Springer International Publishing, Cham, pp 291–308
4. Essletzbichler J (2016) Renewable energy technology and path creation: a multiscalar approach to energy transition in the UK. In: Path dependence and new path creation in renewable energy technologies. Routledge, pp 63–88
5. Gregory J, Geels FW (2024) Unfolding low-carbon reorientation in a declining industry: a contextual analysis of changing company strategies in UK oil refining (1990–2023). Energy Res Soc Sci 107:103345
6. IPCC (2006) Guidelines for national greenhouse gas inventories. In: Eggleston HS, Buendia L, Miwa K, Ngara T, Tanabe K (eds) National Greenhouse Gas Inventories Programme. IGES, Japan
7. Dixon J, Bell K, Brush S (2022) Which way to net zero? A comparative analysis of seven UK 2050 decarbonization pathways. Renew Sustain Energy Trans 2:100016
8. Hammond GP (2022) The UK industrial decarbonization strategy revisited. Proc Instit Civil Eng-Energy 175(1):30–44
9. Kellett J (2003) Renewable energy and the UK planning system. Plan Pract Res 18(4):307–315
10. Alnasrawi A (2001) Iraq: economic sanctions and consequences, 1990–2000. Third World Quart 22(2):205–218
11. Pasha AK (2008) US invasion of Iraq and Indo-Iraq relations. Jadavpur J Int Relat 11(1):181–212

Chapter 5
Carbon Accounting and Footprint Calculation

Abstract Changes in the global climate and increased emissions from preindustrial levels challenge biodiversity and ecosystem services. The increasing energy demand and fossil fuel consumption drive global carbon emissions. This chapter introduces the organization's carbon accounting measures for reducing greenhouse gas emissions. Most companies account for their carbon footprint with the objective of reducing future carbon emissions and taking significant measures for carbon reduction. The carbon accounting procedure considers the Scope 1, Scope 2, and Scope 3 emissions of an organization. Carbon accounting is a complex accounting procedure that requires real-time emission and historical energy data and emission factors. This chapter introduces the procedure of emission accounting with examples of companies such as SSE Energy Solutions and Apple Inc. These illustrations provide insights into why companies do their emission accounting and provide strategies for reducing emissions in each of the scopes. The carbon accounting procedure has gained popularity, as stakeholders such as investors, policymakers, and consumers prefer to account for carbon emissions, with a focus on achieving net-zero emissions and aiming to reduce the risk of their investments.

Keywords Carbon accounting · Carbon footprint · Scope 1,2,3 emissions · GHG protocol · Greenhouse gas emissions

5.1 Introduction

In recent years, companies have increasingly taken measures to reduce their carbon footprint to minimize its environmental impact [4, 8, 16, 19]. Climate change refers to long-term shifts in temperatures and weather patterns, which are caused mainly by human activities, especially the burning of fossil fuels. Climate change is a challenge to sustainability resulting from the key economic drivers of growth [22]. The UNFCCC, Kyoto Protocol 1997, was the first binding treaty on climate change. The Kyoto Protocol paved the way for market-based mechanisms for addressing climate change. Countries focus on achieving the COP28 target of tripling global renewable capacity by 2030 [17]. Organizations and corporations focus on the goals of net-zero emissions. Organizations announce their commitment to achieving net-zero carbon emissions by 2050 through emission accounting. Emission accounting involves quantifying the greenhouse gases that a business emits and, subsequently, their impact on climate change. Carbon accounting is a technique used to understand an organization's carbon emissions [3, 20, 22, 23]. The origin of the carbon footprint is the concept of the ecological footprint proposed by Wackernagel and Rees [26]. According to [9], the carbon footprint refers to the land area required to assimilate the CO_2 produced by human beings. Over the course of time, the use of a carbon footprint is common in a modified form for corporate carbon footprint calculations. The carbon footprint is a quantitative expression of GHG emissions from an economic activity [6, 7]. These GHG data are collected through direct real-time emission data measurement or the use of conversions with emission factors. The carbon footprint is an effective tool for estimating the carbon emissions of businesses or households. Greenhouse gas emissions are the primary driver of climate change. Carbon accounting focuses on quantifying GHG emission data. Emission accounting is a procedure that captures how much GHG a business or organization emits.

5.2 Why Carbon Accounting

Figure 5.1 shows the word cloud of key terms associated with carbon accounting. Corporations experience increasing pressure from consumers, investors, and policymakers to disclose both their direct greenhouse gas (GHG) emissions and supply chain (Scope 3) emissions. There are three sets of stakeholders asking for emission transparency:

- Investors want to reduce emissions and reduce risks.
- Policymakers aim to reduce emissions to achieve net-zero emissions.
- Consumers prefer a reduction in emissions and a clean environment.

Sources of greenhouse gases include carbon dioxide (CO_2), methane (CH_4), nitrous oxide (N2O), hydrofluro carbons (HFCs), perfluorocarbons (PFCs), and sulfide (SF6). While emissions may be the result of any of these gases, the standard unit for measuring emissions is CO_2e (carbon dioxide equivalent). The carbon accounting procedure uses emission factors for conversion.

Carbon accounting is the method of quantifying an organization's greenhouse gas (GHG) emissions. It is also known as the corporate carbon footprint. Organizations use carbon accounting to understand their carbon footprint. It includes emissions from the organization's direct operations and activities, such as heating office buildings, as well as indirect emissions, such as emissions generated by a company's suppliers or by end consumers using its products. The accounting procedure converts GHG

5.2 Why Carbon Accounting

Fig. 5.1 Word cloud of key terms associated with carbon accounting

and CO2 to CO2e. As per the GHG Protocol Corporate Standard,[1] GHGs are classified into three scopes: Scope 1, 2, and 3 emissions.[2] To help delineate direct and indirect emission sources and improve transparency, the Greenhouse Gas Protocol [27] defines three scopes of emissions for reporting purposes [5, 11, 18, 21, 28, 32].

Scope 1 + Scope 2 + Scope 3 = Corporate carbon footprint
(Burn) (Buy) (Beyond).

Scope 1 emissions refer to direct emissions. Examples include buildings and vehicles that the organization directly owns or controls.

Scope 2 emissions are indirect emissions resulting from purchased electricity, heating, steam, and cooling.

Scope 3 emissions refer to all other indirect emissions across the organization's upstream and downstream value chains. Scope 3 emissions occur in the value chain; most carbon emissions are Scope 3 emissions. Upstream activities refer to emissions from the production or extraction of purchased materials, and downstream activities refer to emissions that occur as a consequence of the use of the organization's products or services.[3] There are 15 categories of Scope 3 emissions.[4] They include purchased goods and services, capital goods, fuel and energy-related activities not included in

---

[1] GHG protocol is an initiative of businesses, NGOs, governments, and others convened by World Resources Institute (WRI), US-based Environmental NGO, WBCSD, and Geneva-based 170 international clusters of companies. It was launched in 1998. It is an accounting and reporting standard for businesses. https://ghgprotocol.org/sites/default/files/standards/ghg-protocol-revised.pdf.

[2] https://www.energycap.com/resource/carbon-accounting-101-ebook/

[3] https://www.energycap.com/resource/carbon-accounting-103-ebook/

[4] https://ghgprotocol.org/scope-3-calculation-guidance-2.

Scope 1 or Scope 2, upstream transportation and distribution, waste generated in operations, business travel, employee commuting, upstream leased assets, downstream transportation and distribution, processing of sold products, use of sold products, end-of-life treatment of sold products, downstream leased assets, franchises, and investments. The estimate guidelines are based on the GHG protocol.

The GHG protocol is an initiative of businesses, NGOs, governments, and others convened by the World Resources Institute (WRI), the US-based Environmental NGO, WBCSD, and the Geneva-based 170 international clusters of companies [27]. It was launched in 1998. It is an accounting and reporting standard for businesses. There are two standards for the GHG Protocol. 1. Product Life Cycle Accounting and Reporting Standard. 2. Corporate Accounting and Reporting Standard. These accounting standards provide specific calculation tools for GHG quantification. The carbon footprint has emerged as a strong indicator of GHG emissions. Carbon accounting is essential for organizations to understand the extent of their emissions and helps them take steps to reduce their emissions. Carbon accounting helps capture organizations' direct and indirect emissions; through the reporting of their emissions, they can set emission reduction targets and establish strategies for tracking their emissions. The emission accounting procedure involves the following steps: 1. Measuring emissions. 2 Setting emission reduction targets. 3. Implementing strategies for emission reduction. 4. Reporting their emissions. Reporting can be performed via annual ESG reporting and monthly carbon reporting.

5.3 SSE Energy Solutions Carbon Footprint

The SSE is Scottish, and Southern Energy is one of the largest suppliers of electricity and gas in the UK. The SSE's primary focus is on maintaining electricity grids in the UK, the EV infrastructure, renewable energy generation, and business energy solutions. SSE was initially a Big Six energy supplier. Furthermore, it was sold to a retail energy business (SSE Energy Services). SSE Energy Solutions are businesses that provide low-carbon energy solutions for clients and focus on net-zero solutions to reduce GHG emissions. SSE Energy Solutions provides energy services, including renewable energy solutions, in the UK and Northern Ireland [24, 25, 30, 33]. The following is a breakdown of the Scopes 1, 2, and 3 emissions of the SSE Energy Solutions, as shown in Tables 5.1 and 5.2.

The table above shows the Scope 1, 2, and 3 emissions of SSE Energy Solutions in the UK. Scope 1 emissions, which are directly owned by SSE Energy Solutions, include office emissions (0.6 metric tons of CO2e), and other Scope 1 emissions (0.0 metric tons of CO2 e). Scope 2 emissions, which include office emissions from the acquired sources of energy from outside, are 3.9 metric tons. Scope 3 emission is the emission due to the upstream and downstream activities of the SSE Energy Solutions.

According to the World Resource Institute, 75% of companies' greenhouse gas emissions are Scope 3 emissions (on average). In this example, Scope 3 emissions

5.3 SSE Energy Solutions Carbon Footprint

Table 5.1 Scope 1, 2, and 3 emissions of SSE energy solutions for the years 2020–2021

	Description	Metric tons of CO_2e
Scope 1	Office emissions (Gas)	0.6
	Other Scope 1 emissions	0.0
Scope 2	Office Emissions (Electricity)	3.9
Scope 3	Homeworker electricity emissions (Home office)	3.2
	Homeworker electricity emissions (air-conditioning)	0.0
	Homeworker gas emissions	14.1
	Diesel emissions (commute)	25.2
	Petrol emissions (commute)	34.2
	Hybrid emissions (commute)	5.4
	Motorbike emissions (commute)	1.6
	Train emissions (commute)	1.6
	Bus emissions (commute)	1.9
	Taxi emissions (commute)	0.0
	Train emissions (in work travel)	2.4
	Diesel emissions (in work travel)	2.7
	Petrol emissions (in work travel)	2.8
	Hybrid emissions (in work travel)	0.0
	Bus emissions (in work travel)	0.0
	Taxi emissions (in work travel)	0.0
	Plane emissions	0.6
	Hotel emissions	0.8
	Additional estimates for purchases of goods and services	2.5
	Other scope 3 emissions	0.0

Source SSE energy solutions, report 2020–2021

Table 5.2 Total scope breakdown of SSE energy solutions

SSE	Carbon footprint
Scope 1	0.6
Scope 2	3.9
Scope 3	98.9

include homeworker electricity emissions, emissions from commuting, business travel, and other Scope 3 emissions.

The breakdown of Scope 1, Scope 2, and Scope 3 emissions is shown in the above table; the total carbon footprint is 103.4 tons of CO_2e. The SSE carbon footprint calculations have implications for corporate sustainability strategies and achieving net-zero targets. Tracking Scope 1, 2, and 3 emissions will allow the SSE to track carbon emissions. This will allow the SSE to refine its renewable energy expansion plans and substantially contribute to decarbonizing the UK's energy systems. Accounting for emissions will help reduce emissions in various Scope 1, 2, and 3 SSE Energy Solutions, including the supply chain, and pave the way for offsetting measures. Stakeholders such as investors and consumers will enhance their market reputation, quickening the UK's transition to net-zero emissions [10, 12].

Offsetting Measures

1. Million Trees Pledge

The Million Trees Pledge is SSE's broader commitment to address climate change. It attracts businesses, organizations, and individuals to offset their emissions. Through the Million Trees Pledge, businesses commit to planting 1 million trees over a period of 10 years. This combats climate change by absorbing CO_2 emissions, thus supporting biodiversity through afforestation. Through these efforts, 10 + million trees were planted, and 51 + million trees were pledged [29].

2. Investment in Renewable Energy Projects

The SSE has a substantial focus on renewable energy projects such as solar, wind, and battery storage (Dogger Bank Wind Farm, Seagreen Offshore Wind Farm, and Galway Wind Park). They are integral to the sustainability goals and achieve net-zero emissions by 2050.

5.4 Apple's Strategies for Carbon Footprint Reduction

Apple aims to achieve carbon neutrality through its operations and supply chain by 2030. Apple's commitment to achieving 100% carbon neutrality for its supply chain ensures that every piece of Apple equipment has a net-zero effect [31]. Apple's roadmap for carbon neutrality provides guidance to other industries to reduce their carbon footprint. Apple aims for innovative solutions to reduce its carbon footprint and set a benchmark for global emission reduction targets. As a major global corporation, Apple's actions may influence energy trading dynamics and carbon pricing, especially in regions where they operate. Apple has made significant strides in its climate action. Apple 2030 showed its commitment to becoming carbon neutral, eliminating emissions from its supply chain. Compared with the base year 2015, Apple has reduced its overall greenhouse gas emissions across Scopes 1, 2, and 3 by more than 55% [34, 35]. The following includes Apple's decarbonization strategies:

1. Materials: Apple prioritizes materials that have a lower carbon footprint.
2. Manufacturing: Apples use recycled content for their products, such as the Macbook, and reduce emissions from manufacturing processes. Apple encourages suppliers to adopt sustainable practices and reduce their carbon footprint.
3. Recycling: MacBook Air is the first Apple product made with 50% recycled content.
4. Renewable Energy: It is estimated that Apple's environmental programs avoided 31 million metric tons of emissions across all scopes in 2023. Initiatives that we have been growing for years continue to yield clear results, including sourcing 100% renewable energy for our facilities, transitioning suppliers to renewable energy, and using low-carbon materials in products.

5.4 Apple's Strategies for Carbon Footprint Reduction

Apple invests in innovative technologies to achieve the targets of sustainability. Through its efforts to achieve sustainability, Apple reduces carbon emissions. Through its innovative approach to new materials, manufacturing, recycling, and advanced battery technology, Apple has adopted significant strategies to make significant progress in reducing carbon emissions in the tech industry. Apple aims for 100% recycled and renewable energy by 2030. Apple has created new recycling technologies to recover useful materials. Apple has reduced mining waste and reused materials such as aluminum in the production of electronic devices with the objective of decarbonization [2]. Apple's green transition includes innovations, regulatory compliance, and targets for global sustainability goals. Apple shows proactive leadership in its renewable energy transition achieving carbon neutrality. Apple's engagement in clean energy strategies is aligned with international agreements, and Apple advocates strong policies for climate mitigation. Tech companies such as Google, Microsoft, and Amazon take specific steps to reduce their carbon footprint by using less energy and recycled materials. Google adopted 100% renewable energy in its operation centers in 2017 [13–15]. Microsoft and Amazon are committed to removing carbon emissions via sustainable energy sources. Amazon is targeting net-zero emissions by 2040 with major investments in green energy [1].

Summary

This chapter introduces carbon accounting as a corporate strategy for emission reduction, the procedure of emission accounting, and Scope 1, Scope 2, and Scope 3 emissions. Carbon accounting is a global priority for businesses and organizations, as carbon emissions are increasing rapidly. Research on carbon accounting is new; businesses quantify their carbon emissions to reduce carbon emissions and seek offsetting strategies. The procedure of carbon accounting is illustrated with the example of SSE Energy Solutions' report on Scope 1, Scope 2, and Scope 3 emissions. Insights from Apple's environment report are discussed in this chapter. Accounting for greenhouse gas emissions is inevitable for controlling future emissions and finding strategies for offsetting emissions. Sustainable business practices are crucial for addressing global climate change. New strategies for accounting for emissions to mitigate climate change are at the forefront of international debates. As businesses strive to reduce their carbon footprint, they deliberate their commitment to a green economy. Thus, carbon accounting has become a dispensable part of mitigating greenhouse gas emissions.

Questions

1. What outcomes can be observed from Apple's sustainability strategies, and how do these contribute to shaping the global energy market and achieving the targets of Net Zero?

References

1. Amazon (2024) The climate pledge. Amazon Climate Pledge
2. Apple Inc's (2022) Apple expands the use of recycled materials across its products. https://www.apple.com/environment/
3. Ascui F (2014) A review of carbon accounting in the social and environmental accounting literature: what can it contribute to the debate? Soc Environ Account J 34(1):6–28
4. Ascui F, Lovell H (2011) As frames collide: making sense of carbon accounting. Account Audit Accountabil J 24(8):978–999
5. Bowen F, Wittneben B (2011) Carbon accounting: negotiating accuracy, consistency and certainty across organisational fields. Account Audit Accountabil J 24(8):1022–1036
6. Carbon Trust (2007a). Carbon footprint measurement methodology, version 1.1. The Carbon Trust, London, UK. http://www.carbontrust.co.uk. Accessed on 27 Feb 2008
7. Carbon Trust (2007b) Carbon footprinting. An introduction for organizations. http://www.carbontrust.co.uk/publications/publicationdetail.htm?productid=CTV033. Accessed on 5 May 2008
8. Dasgupta PS, Heal GM (1979) Economic theory and exhaustible resources. Cambridge University Press
9. East AJ (2008) What is a carbon footprint? An overview of definitions and methodologies. In: Vegetable industry carbon footprint scoping study—discussion papers and workshop. Horticulture Australia Limited, Sydney
10. Elliott D (2016) Renewable energy in the UK: a slow transition. Green economy reader: lectures in ecological economics and sustainability. Springer International Publishing, Cham, pp 291–308
11. Energy Systems Catapult (2023) Carbon accounting and standards in industry: a framework for innovation and growth
12. Essletzbichler J (2016) Renewable energy technology and path creation: a multiscalar approach to energy transition in the UK. In: Path dependence and new path creation in renewable energy technologies. Routledge, pp 63–88
13. Google (2016) Accelerating climate action with AI. https://blog.google/outreach-initiatives/sustainability/report-aisustainability-google-cop28/
14. Google (2024a) 100% renewable is just the beginning. 100% Renewable Energy Projects at Google
15. Google (2024b). Operating on 24/7 carbon-free energy by 2030. Tracking our carbon-free energy progress—google sustainability
16. Green J, Newman P (2022) Carbon accounting: a systematic literature review and directions for future research. Green Finance 4(2):1–22
17. He R, Luo L, Shamsuddin A, Tang Q (2021) Corporate carbon accounting: a literature review of carbon accounting research from the kyoto protocol to the Paris agreement. Accounting & Finance
18. IPCC (2006) Guidelines for national greenhouse gas inventories. In: Eggleston HS, Buendia L, Miwa K, Ngara T, Tanabe K (eds) National greenhouse gas inventories programme. IGES, Japan
19. Mahapatra SK, Schoenherr T, Jayaram J (2021) An assessment of factors contributing to firms' carbon footprint reduction efforts. Int J Prod Econ 235:108073
20. Marlowe J, Clarke A (2022) Carbon accounting: a systematic literature review and directions for future research. Green Finance 4(1):71–87
21. Pandey D, Agrawal M, Pandey JS (2011) Carbon footprint: current methods of estimation. Environ Monit Assess 178:135–160
22. Schaltegger S, Csutora M (2012) Carbon accounting for sustainability and management. Status quo and challenges. J Clean Prod 36:1–16
23. Stechemesser K, Guenther E (2012) Carbon accounting: a systematic literature review. J Clean Prod 36:17–38

References

24. SSE Energy Solutions (2021) Carbon footprint report, 2020–21
25. SSE Sustainability Report (2024)
26. Wackernagel M, Rees W (2014) What is an ecological footprint? In: From our ecological footprint (1996). In sustainable urban development reader. Routledge, pp 375–383
27. World Resources Institute & World Business Council for Sustainable Development (2004) Chapter 4: Setting organizational boundaries. In: The greenhouse gas protocol: a corporate accounting and reporting standard. World Resources Institute, pp 16–24
28. https://ghgprotocol.org/sites/default/files/standards/ghg-protocol-revised.pdf
29. https://www.milliontreepledge.org/
30. https://www.gov.uk/government/publications/greenhouse-gas-reporting-conversion-factors-2020
31. https://www.apple.com/environment/pdf/Apple_Environmental_Progress_Report_2024.pdf
32. https://www.energycap.com/resource/carbon-accounting-101-ebook/
33. https://www.sseenergysolutions.co.uk/customer-help-centre/help-and-advice/streamlined-energy-and-carbon-reporting
34. https://www.apple.com/environment/pdf/Apple_Environmental_Progress_Report_2021.pdf
35. https://www.apple.com/environment/pdf/Apple_Environmental_Progress_Report_2023.pdf

Chapter 6
Renewable Energy Transition and Energy Policies

Abstract The UK's energy system is based on fossil fuels. However, in recent years, renewable energy has become a part of the transition to a sustainable energy model. Sustainable livelihoods are a priority for market economy, governments, and the public. The U.K. has numerous renewable energy sources, including onshore and offshore wind farms, solar housing projects, and tidal energy initiatives, to achieve its net-zero targets by 2050. This chapter discusses emerging examples of sustainable community models of renewable energy projects. In this chapter, we discuss the Hockerton Housing Project, an example of a sustainable community and livelihood, and the Hornsea offshore wind farm, which is the largest wind farm in the world and the energy policies from a global context. Sustainable livelihoods and practices for renewable energy adoption increase energy efficiency, reduce energy costs, and reduce carbon emissions. Community renewable energy projects have played a crucial role in transitioning the energy system in the UK. This chapter examines the renewable energy transition in the UK, energy policy, and politics of renewable energy.

Keywords Renewable energy · Sustainable livelihood · UK's carbon footprint · Energy policy

European countries are pioneers in achieving renewable energy targets and net-zero emissions. The European Union achieved the target of supplying 20% of energy from renewables in 2020 and 27% in 2030 [3]. Free market proposals have significantly influenced renewable energy-oriented research and development in the UK since the oil shocks of the 1970s. Free market environmentalism with fewer state-owned enterprises takes deliberate steps to protect the environment more effectively. The UK has been reducing coal since the 1960s, with nearly 141 coal mines in the 1980s, nearly 70% of which were decommissioned, and the UK signed the Kyoto Protocol at the end of the 1990s, further entering into the European carbon market in the 2000s.

Approximately 80% of UK emissions are from fossil fuel emissions, and less than 20% are from land use, waste, and agriculture in 2023. By adopting the carbon plan in 2009, the UK set the goal of generating 30% of electricity from renewable energy in 2020 and made investments of 120 billion GBP in offshore wind farms. The UK increased its renewable energy generation with its solar adoption and offshore wind farms [1, 2, 8, 17, 18, 28]. The share of renewable energy, such as solar, wind, and hydropower, in power generation has increased to 40% by 2022. This reduced coal-based power generation and carbon emissions in the UK. The UK's net-zero strategies highlight community participation. Local authorities, businesses, and households integrate renewables into energy transition efforts. The role of communities in energy transition and sustainable development is significant in the UK's approach to a low-carbon pathway. The UK's Low Carbon Transition Plan (2009) encourages community participation in mitigating climate change [11, 19]. The UK government's community energy strategy highlights that community-led action can address challenges and meet local needs with the support of the local population [4–6, 25].

6.1 Hockerton Housing Project

The Hockerton project is in Northamptonshire and was established in the early 1990s. It is a newly constructed development on 18 acres of land: Land with five sustainable homes that use solar energy for their domestic needs. Earth-sheltered homes have a high thermal mass and high degree of insulation. These five households utilize sustainable energy sources, grow fruits and vegetables, and recycle waste. This is one example of a sustainable livelihood. A community-owned wind turbine is a cost-effective method to offset carbon emissions. The Hockerton Housing Project is an example of sustainable community living. It is an earth-sheltered environmental housing project. Residents of these five houses generate clean energy and practice a sustainable living environment, preventing emissions [28]. DETRs funded a part of this project.

Community initiatives are a sustainable model of decentralized energy generation. The Hockerton Housing Project (HHP) is an example of a sustainable community model. Community-owned energy generation ensures cost reduction, energy efficiency, and sustainability. This is a localized approach to address climate change and environmental challenges. Community participation is essential to achieving sustainable outcomes in rural–urban settlements. Five households in the HPP collectively manage their energy generation (see Fig. 6.1). The HPP supplies locally generated electricity to the grid. Solar and wind turbines' energy sources offer income and reduce the cost of energy used in their households.

Rainwater harvesting, solar and wind energy generation, and community fruit and vegetable production constitute ways to achieve sustainable living. Community-led energy projects [23] have reduced the dependence on fossil fuels and offered a sustainable livelihood. In this sustainability project, the community manages a local energy project. This is an example of how the collective efforts of the rural community can

6.2 Hornsea Offshore Wind Farm Projects (Hornsea 1 and 2)

Fig. 6.1 Field visit, Hockerton housing project, Nottinghamshire

contribute to sustainability and emission reduction through community participation. Another example of sustainable urban living is BedZED.

Beddington Zero Energy Development (BedZED) in Sutton, South London, is a sustainable, low-carbon, environmentally friendly urban living model. This is one example of the UK's efforts to sustain energy-efficient homes through sustainable innovation-led urban development. BedZED started in 1997 through collaboration between Bioregional and architect Bill Dunster. The Sutton Borough Council utilized unused land to develop a low-cost, low-energy housing project that benefits households. This project shows how collaboration among builders, developers, architects, and local governments facilitates sustainability-led urban housing.

6.2 Hornsea Offshore Wind Farm Projects (Hornsea 1 and 2)

The UK has onshore and offshore wind farms promoting the renewable energy transition. The Hornsea project is situated in the North Sea, strengthening the UK's decarbonization and energy security. This project includes Hornsea 1 and 2, developed by Orsted, aligning the goals of decarbonization and achieving the net-zero targets. Hornsea 1 is the largest offshore wind farm in the world and was previously owned by DONG, a Danish energy company. Hornsea 1 was established in 2020 and is one of the milestones of the UK in its renewable energy development. It offers electricity to a million households in the UK and fosters the renewable energy transition in the UK. Orsted has adopted a hybrid financing model through debt financing and joint ventures, and the revenue is reinvested in significant renewable energy projects. Orsted agreed with long-term power purchase agreements (PPAs) with energy buyers.

Orsted developed Hornsea 2 in 2022 with a capacity of 1.32 GW and 165 wind turbines, which are suitable for supplying electricity to approximately 1.5 million households in the UK. Hornsea projects can achieve 50 GW of offshore wind capacity in 2030. These projects indicate the UK's significant role in global renewable energy generation. Wind farm projects work under the Contract for Differences (CfD), ensuring fixed electricity prices from low-carbon sources. Hornsea projects are flagship projects of the UK's offshore wind deployment and sustainable innovation. The development of these projects has created external economies, generating employment and regional economic growth in the coastal regions of Yorkshire. This project will increase the UK's renewable energy infrastructure and clean energy capacity. The Hockerton Housing Project (HHP) and BedZED are community-based examples that integrate environmental sustainability. This community-based model offers a model for the growth of sustainable cities. These models set a benchmark for a sustainable city, but their restricted scale indicates that the expansion of a sustainable city is a long way.

Despite the UK's transition to renewable energy, several challenges are posed by the intermittency and variability of weather events. Wind and solar energy are highly dependent on weather events. This creates a demand–supply mismatch during seasons with less sunshine and wind [20]. In the rural areas of Scotland and Wales, the renewable potential is high, but there are potential grid constraints and connection capacities that increase costs [22]. Although the cost of renewable energy projects is relatively low, the largest scale infrastructure requires upfront costs. The UK has made significant strides in renewable energy. However, practical limitations must be addressed through investment, sustainable innovation, and policies.

6.3 UK's Carbon Footprint

Figure 6.2 explains the extent of emissions from electricity generation, consumer expenditure, manufacturing, transport, and other sectors during 1990–2023. Greenhouse gas emissions from electricity generation significantly decreased from 218 Mt CO2e in 1990 to 70 Mt CO2e in 2023. Emissions from manufacturing decreased from 218 Mt CO2e in 1990 to 73 Mt CO2e in 2023, reflecting energy efficiency. Emissions from the transport sector were relatively modest, at 67 MT CO2e in 1990, and increased to 84 Mt CO2e in 2023. Emissions from other sectors, such as agriculture,

6.3 UK's Carbon Footprint

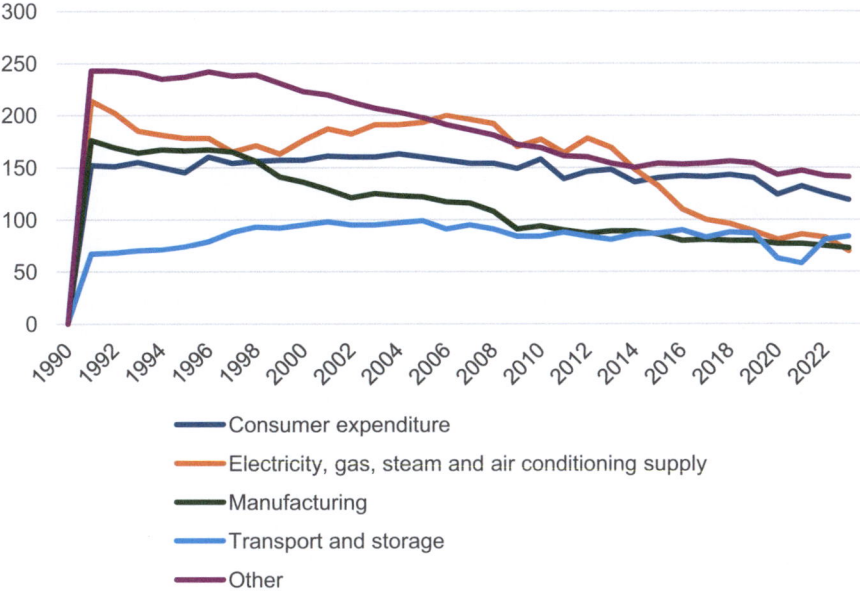

Fig. 6.2 Major emitting industries on a residence basis: UK, 1990 to 2023. Greenhouse gas emissions, UK: provisional estimates from the Office for National Statistics (ONS) (in million tonnes of CO2e)

industry, land use, and waste, are 141 Mt CO2e in 2023, the highest among all emission categories. The decline in emissions from the electricity and manufacturing sectors implies the UK's successful decarbonization strategies for the low-carbon energy transition [27].

The UK has made substantial progress in some industries, but other sectors, such as waste, land use, and agriculture, have high emissions, necessitating policy interventions to target net-zero emissions by 2050. The technologies used for the low-carbon agenda are PV panels, heat pumps, district heating, etc. In 2021, the UK's footprint emissions were 705 Mt CO2e, which represents a 36% reduction since 1990.[1] The data show that the UK's carbon footprint rose by 15% between 2020 and 2021, mainly due to imported goods, but it peaked at 963 Mt CO2e in 2007. A study by Leeds University revealed that emissions from imports are more than double the 1996 levels. Since 1997, the UK economy has moved to a service-oriented economy. As a result, the UK imports significant consumer goods from overseas; these products are imported from countries such as China, the U.S., the EU, and other countries. The carbon footprint of England was 807 Mt CO2e in 2004, and it was 27% lower in 2021. These findings indicate that England's carbon dioxide footprint rose by 13%

---

[1] https://www.ons.gov.uk/economy/environmentalaccounts/methodologies/measuringukgreenhousegasemissions.

between 2020 and 2021.[2] Half of the world's energy consumption is from heat, 40% of which is energy-related carbon emissions [26]. In the EU, heating accounts for 36% of total emissions, 29% in the U.S., and up to 50% in China. Heat is a primary global target for mitigating climate change [9, 12, 13, 16].

The key factor of the UK's emission reduction strategy is transforming the power sector. Renewable energy projects, such as solar and wind farms, are remarkable in this transition. Solar and Wind power are large-scale renewable energy sources for the power supply to the National Grid. Community-oriented renewable energy projects are models of sustainable energy transitions, but these models in isolation do not yield adequate results. The models of sustainable community projects should be expanded to include the largest number of households to develop a sustainable city. It will establish a benchmark for renewable energy transitions to a low-carbon transition. The UK's growing trend in the renewable energy transition is via the participation of local authorities and communities. Integration of councils, businesses, and people will strengthen collective efforts to mitigate climate change. The UK has taken significant steps to reduce the carbon footprint of various sectors, such as power, transport, and other sectors. The success of these measures is entirely dependent on the collective action of the communities. The UK employs regulatory measures such as the UK ETS to reduce greenhouse gas emissions, setting a cap on the highest emissions.

6.4 Energy Policies and Recent Developments

Each country has developed feasible strategies for domestic energy generation [14, 24]. The energy policy of any country depends on several scenarios and its vision for innovation and technological capabilities [7, 10]. As the EU plans for strategies of decarbonization by 2030 and achieving the goals of carbon neutrality by 2050, it focuses on the European Green Deal as a new development pathway to renewable energy and has adopted the same approach in 2019. This aims for the EU to achieve net-zero emissions by 2050 [15]. The primary source of energy in Poland is coal [21]. Poland is a transition economy in Europe. Decarbonizing Poland's economy is a challenging scenario. Like the former Eastern bloc, Poland does not have a nuclear power plant, such as Bulgaria, the Czech Republic, Hungary, Romania, and Slovakia [15]. Therefore, to manage this scenario, Poland would cooperate with neighboring nuclear power producers such as the Czech Republic, Belarus, Slovakia, Germany, and Romania through the power grid. Most nuclear power plants have a significant capacity to produce energy, but they produce up to 80% to 90% of their capacity. Through cooperation, Poland can buy energy from these countries through the power grid. Poland tries to establish a nuclear power plant with the support of Japan. Most nuclear power plants have a high capacity, but their actual energy production is significantly less than their production capacity. Increasing the production capacity

---

[2] https://www.gov.uk/government/statistics/uks-carbon-footprint/carbon-footprint-for-the-uk-and-england-to-2019.

by two to three percent from various nuclear power plants in neighboring countries of Poland is a suitable method for Poland to mitigate the energy crisis.

Establishing a new nuclear power plant in Poland is an expensive long-term investment. It also has significant environmental concerns in the future. Thorium-based power plants are considered better options for green energy than uranium-based nuclear-based power plants. As of 2024, Poland established its first gas-based power plant. This is a significant step to avoid coal-based energy production. However, this has become expensive as the Ukraine war prevails. Germany had a plan to decommission its nuclear power plants before the Ukraine and Russia war. They planned to establish new gas-based electricity from cheaper supply options in Russia. After the Ukraine war, Germany postponed its nuclear power plant decommissioning. Therefore, Poland must look for alternative options for clean energy.

Energy crisis will be a problem for Poland if Poland adopts a similar global manufacturing production strategy as China did. This production strategy is beneficial for Poland to improve its domestic manufacturers' capacity in the area of FMCGs and machinery equipment sector in Poland. Production should focus on emerging countries such as Europe and Global South in the globalized world to foster the benefits of the strategic production of transitioning Poland's economy from a domestic producer economy to an export-led economy. This might benefit the production of energy from the establishment of nuclear or thorium power plants.

Poland looks forward to more FDI investments from Japan and China in technology and innovation. It needs further investment in high-speed railway machinery and equipment and vehicle manufacturing, and it requires more IT investments from India. China and Poland will have to cooperate on refinery establishments in Poland. This will lead to better trade negotiations with other EU countries. The anticipation of energy security will provide suitable methods for addressing alternative green energy sources in Poland. At this juncture, it is crucial to accepting biogas energy investments from Germany and China. This will help to adopt a transition in Poland's energy mix and navigate to net-zero emissions in 2050.

6.5 India's Energy Policy

Fossil fuels dominate the global energy demand, and crude oil is the largest energy source globally, accounting for 31.6% of the total energy supply, and coal accounts for 26.9% [12]. One of the biggest goals of developed countries like the UK is to achieve a renewable energy transition and net-zero emissions by 2050. Coal is the second largest source of energy in the world, contributing 70% of India's electricity generation. Coal production and consumption are high in India. The Union Budget 2025 marked a major shift toward India's renewable energy transition, with the adoption of nuclear energy in the energy mix. Attracting private sector participation in nuclear energy—the transition to renewable energy—highlights the importance of nuclear energy capitalism. The initiative underscores the recognition of nuclear power for India's energy integration, but it needs to be carefully considered. Nuclear energy is full of challenges due to safety concerns and the dangerous nature of radioactive materials. India is a highly corrupted and populated country, and nuclear capitalism will lead to environmental issues in the future.

Recent Union Budget of India set a highly ambitious transition of the energy mix, through the adoption of nuclear energy. France has the highest amount of nuclear power, and developed countries such as Germany and the Netherlands have made more investments in nuclear energy. Despite technological developments, Japan has experienced catastrophic nuclear disasters. The burden of a nuclear disaster is always on the lives of the people and the environment. While nuclear energy has a low-carbon strategy for fossil fuels, its safety risk makes it an impractical choice for India's energy transition. Instead of leaning toward nuclear power, India should double-down on renewables such as solar or wind farms alongside energy efficiency measures to foster the green transition. Given the scarcity of fossil fuels and their harmful emissions, India should look for feasible alternatives. India has significant wind potential; using wind as a promising renewable energy source is suitable. Unlike the cold weather regions of the UK and Europe, India has potential opportunities for solar adoption. The Indian administration is considering private nuclear companies for investment in India. This may lead to mere profitable gains for political parties while risking the hazardous nature of nuclear energy. There are various options for adopting renewable energy in India rather than investing in a nuclear energy-led renewable transition; nuclear companies wrongly consider it a sustainable energy resource because of their campaigns through different sustainability organizations.

Ironically, India wants to reduce the solar dependability of panels from China; China has higher capabilities than India in terms of renewable energy. India tries to associate with Western nuclear companies to reduce Chinese influence to obtain world support for profit motives. The liberal investment policy of the nuclear energy market in India has led to greater investment opportunities from different private companies worldwide. Investors have greater investment capabilities, which leads to uncontrollable corruption with the support of the government system in India. The political parties use these potential benefits to counter their opposition. Nuclear companies can act as they wish to increase their investment capabilities. This has led to a decline in the quality of building nuclear power plants and nuclear waste management. In developing countries such as India, private companies do not adhere to the same standards they use in developed nations. Many developed nations are trying to shut down their nuclear plants. Italy does not have any nuclear power plants because of environmental concerns. India has the largest population in the world and the most prominent migrant workforce. India's nuclear power transition is unsuitable for sustainability goals.

An effective administration should reject nuclear liberalization and adopt green energy through environmentally friendly methods such as biogas. China is the world's largest producer of biogas. India should increase its FDI so that Chinese companies can invest in green energy and innovation. A lack of innovation in green technology leads to nuclear liberalization. India should invest more innovation funds in private enterprises, government institutions, and universities for innovations in green energy. Nuclear capitalism will let India stop other alternatives to green energy. Nuclear energy is a long-term disaster for the environment. Disasters caused by nuclear plants lead to increased expenses to counter the negative impacts of radiation and other problems in the environment.

The West and the Global South should consider other green energy investment opportunities rather than nuclear energy. The green energy transition is possible only through private and government cooperation. India should open its market for Chinese and Western companies to invest in biogas, wind energy, solar energy, etc. This opportunity will bring sustainable energy to India without risking people or the environment. The transition to green energy is possible only through investing in natural green energy resources, not nuclear energy. India should open markets for Western and Chinese companies to invest in green energy resources by adopting liberal green energy policies such as subsidies and zero taxation. This will lead to a better green energy transition to India rather than the wrong policy adopted for nuclear energy. China has established a fourth-generation nuclear plan, while the U.S. is beginning to do so soon. Moreover, private players are eager to invest in India at a significantly lower cost, especially outdated nuclear power plants. This will prove detrimental in the long run.

6.6 Politics of Renewable Energy

The road to renewable energy is often associated with investments in solar, wind, tidal, and hydropower. To mitigate the climate crisis, we must reserve renewable energy for current and future years. The industrial growth and energy market of every economy are threatened by climate injustice. On the one hand, the concentration of fossils impacts industrial production; on the other hand, CO_2 emissions increase. To address climate change, countries must take responsibility for reducing CO_2 emissions. To address the climate crisis, renewable energy companies have greater influence on global politics. Most renewable energy technology is controlled by Western countries. The increase in renewable energy in China is different from that in the Global South. West China and the United States have sources of natural gas-fueled renewable energy, which is a less CO_2-emitting source and safer than nuclear power.

After Russia–Ukraine, electricity generation via natural gas is an expensive option compared with nuclear power, and Europe has fewer alternatives than does the adoption of renewable energy. Owing to its geography and climate, options are available for adopting 100% renewable energy production capabilities. The alternative options of the EU and the U.S. include nuclear power plants, and biogas. Compared with fossil fuel sources, nuclear power plants cause less pollution. The EU has a smaller population than the Global South does; this may be a comparative advantage in establishing a nuclear power plant for renewable energy generation. Technological capabilities and capital investments are important for the safety of nuclear power plants, which is a possibility for the EU. Thorium-based power plants will be a greater advantage for renewable energy generation; Western countries must invest in thorium power plants for sustainable alternatives.

America holds a greater comparative advantage than the EU in terms of renewable energy production supported by favourable climatic conditions and abundant natural resources. While the EU has aggressively invested in renewable technologies to offset declining fossil fuel use, the U.S. still relies heavily on traditional energy sources. Following global negotiations and commitments to adopt market mechanisms, taxing domestic fossil fuel companies in the U.S. would be an encouraging step toward reducing emissions. However, as a classic free-market economy, the U.S. energy transition depends on investment incentives. Returns on investment from conventional energy remain higher than those from nonconventional sources, making renewable investments less attractive is essential for moving investment away from fossil fuels and supporting the country's energy transition. The strategic oil reserves of the U.S. are a major factor in their energy security and an essential element of their oil diplomacy. Therefore, the possibility of the U.S. adopting renewable energy solely depends on the possibility of transitioning American energy companies (Exxon mobl) to renewable energy (for example, transitioning DONG to Orsted). Energy companies, such as DONG, have undergone a 100% transition to renewable energy and have established onshore and offshore wind farms in the UK. Nations with clear vision and sustained commitment, can successfully strengthen their renewable energy markets and move toward a more sustainable energy future.

Summary

Renewable energy is a strategy for reducing the over-dependence on fossil fuels. The transition to renewable energy is a significant policy adoption for many countries. As per state security concerns related to renewable energy, there are some limitations on military and political interests. Fully transitioning to renewable energy is not possible, but some countries can reduce their fossil fuel consumption through renewable energy. Changing from fossil fuel to nuclear power has a serious political dependency on those who don't have nuclear power energy technology. The dominant states will control the nuclear power plants and technology to achieve their interest through the implementation of nuclear power plants. The countries should take a balanced strategy of renewable energy and fossil fuel. Nuclear energy is more suitable for high income countries and less populated countries. Nuclear power plants are not suitable for developing countries due to higher investment and fewer technological innovations. Technology dependency related to nuclear power plants has become a security problem for developing countries.

References

1. BEIS (2022) UK energy in brief 2022. Department for Business, Energy & Industrial Strategy
2. Climate Change Act (2019)

References

3. Cowell R, Ellis G, Sherry-Brennan F, Strachan PA, Toke D (2017) Energy transitions, subnational government and regime flexibility: how has devolution in the United Kingdom affected renewable energy development? Energy Res Soc Sci 23:169–181
4. DECC (2014) Community energy in the UK: Part 2: final report. DECC. https://www.gov.uk/government/uploads/system/uploads/attachment_data/file/274571/Community_Energy_in_the_UK_part_2_.pdf
5. DECC (2015a) Changes to renewables subsidies. https://www.gov.uk/government/news/changes-to-renewables-subsidies
6. DECC (2015b) Community energy strategy update. https://www.gov.uk/government/uploads/system/uploads/attachment_data/file/275163/20140126Community_Energy_Strategy.pdf
7. Edmonts J, Wilson T, Wise M, Weyant M (2006) Electrification of the economy and CO_2 emissions mitigation. Environ Econ Policy Stud 7:175–203
8. Elliott D (2016) Renewable energy in the UK: a slow transition. Green economy reader: lectures in ecological economics and sustainability. Springer International Publishing, Cham, pp 291–308
9. EU (2019) Sustainable buildings for Europe's climate-neutral future. https://ec.europa.eu/easme/en/news/sustainable-buildings-europe-s-climateneutral-future. Accessed on 02 Sep 2020
10. Gabriel J (2014) A scientific enquiry into the future. Eur J Futur Res 2:1–9
11. HM Government (2009) Five point plan the UK Government has a five point plan. https://assets.publishing.service.gov.uk/government/uploads/system/uploads/attachment_data/file/228752/9780108508394.pdf
12. IEA (2017) Global status report 2017. https://www.worldgbc.org/news-media/global-status-report-2017. Accessed on Sept 2020
13. International Energy Agency (IEA) (2018) Market report series renewables 2018 analysis and forecast to 2023. Int Energy Agency 211
14. Kraan O, Chappin E, Kramer GJ, Nikolic I (2019) The influence of the energy transition on the significance of key energy metrics. Renew Sustain Energy Rev 111:215–223
15. Kochanek E (2021) Evaluation of energy transition scenarios in Poland. Energies 14(19):6058
16. Leung J (2018) Decarbonizing US buildings. Center for Climate and Energy Solutions
17. Mitchell C, Connor P (2004) Renewable energy policy in the UK 1990–2003. Energy Policy 32:1935–1947
18. Mitchell C, Watson J, Whiting J (2020) Renewable energy in the UK: past, present and future. Palgrave Macmillan, London
19. Mirzania P, Ford A, Andrews D, Ofori G, Maidment G (2019) The impact of policy changes: the opportunities of community renewable energy projects in the UK and the barriers they face. Energy Policy 129:1282–1296
20. National Grid ESO (2023) Future energy scenarios. https://www.nationalgrideso.com/future-energy/future-energy-scenarios
21. Neutralna Emisyjnie Polska 2050 (2020) Raport. McKinsey & Company, Warszawa, Poland
22. OFGEM (2022) Decarbonization action plan. Office of Gas and Electricity Markets
23. Seyfang G, Park JJ, Smith A (2013) A thousand flowers blooming? An examination of community energy in the UK. Energy Policy 61:977–989
24. Snoek M (2003) The use and methodology of scenario making. Eur J Teach Educ 26:9–19
25. Walker G, Devine-Wright P, Hunter S, High H, Evans B (2007) Trust and community: exploring the meanings, contexts and dynamics of community renewable energy. Energy Policy 38(6):2655–2666
26. Wang Y, Wang J, He W (2022) Development of efficient, flexible and affordable heat pumps for supporting heat and power decarbonization in the UK and beyond: review and perspectives. Renew Sustain Energy Rev 154:111747
27. https://www.buildingcentre.co.uk/media/_file/pdf/22220_pdf29.pdf
28. https://www.hockertonhousingproject.org.uk/

Index

A
Apple's carbon footprint, 56

B
Brent crude oil price, 14

C
Carbon accounting, 51–54, 57
Carbon footprint, 2, 51–57, 64–66
Carbon tax, 5, 6
Climate change, 1, 2, 5, 6, 8, 40, 51, 52, 56, 57, 62, 66, 69
Crude oil exports, 15–18

E
Energy crisis in europe, 13
Energy market, 8, 11, 16, 29, 30, 34, 35, 68, 70
Energy mix, 45, 67, 68
Energy policy, 61, 66, 67, 69
Energy prices, 13, 31, 35, 40, 45
EU ETS, 6, 7

F
Fossil fuels, 1–3, 5, 6, 8, 11–13, 15, 39, 40, 42, 45, 46, 49, 51, 61, 67–70

G
Geopolitics of oil, 18, 21, 24, 25

Global energy demand, 1, 3, 5, 11, 13, 15, 18, 40, 67
Global energy supply, 14
Global production externalities, 1, 2, 8

H
Hotelling rule, 39, 40, 49

N
Negative externalities, 1–3, 5, 6, 8

O
Office emissions, 54
OPEC, 15, 18, 24, 26–28, 30, 31

P
Politics of renewable energy, 61, 69

R
Renewable energy, 13, 39, 40, 42, 45, 46, 48, 49, 54, 56, 57, 61–64, 66–70
Russia–Ukraine war, 2, 14, 15, 18, 31, 45

S
Scope 1, 51, 53–55, 57
Scope 2, 51, 53–55, 57
Scope 3, 51–55, 57
Social cost of war emissions, 1, 3

Sustainable livelihoods, 61, 62

T
Trade war and tariffs, 18, 32, 34

U
UK's carbon footprint, 64, 65

W
War emissions, 1, 3
Wind farms, 42, 56, 61–64, 66, 68

MIX
Papier aus verantwortungsvollen Quellen
Paper from responsible sources
FSC® C105338

If you have any concerns about our products,
you can contact us on
ProductSafety@springernature.com

In case Publisher is established outside the EU,
the EU authorized representative is:
**Springer Nature Customer Service Center GmbH
Europaplatz 3, 69115 Heidelberg, Germany**

Printed by Libri Plureos GmbH
in Hamburg, Germany